Rulers, Guns, and Money

Rulers, Guns, and Money

*The Global Arms Trade
in the Age of Imperialism*

Jonathan A. Grant

HARVARD UNIVERSITY PRESS

Cambridge, Massachusetts

London, England

2007

Copyright © 2007 by the President and Fellows of Harvard College
All rights reserved
Printed in the United States of America

Cataloging-in-Publication Data is available from the Library of Congress
Library of Congress catalog card number: 2006050904
ISBN-13: 978-0-674-02442-7 (alk. paper)
ISBN-10: 0-674-02442-7 (alk. paper)

For Lynn, the love of my life

Contents

Tables

Acknowledgments

This book would not have been possible without the support of many people. I would like to thank the Council on Research and Creativity and the History Department at Florida State University for providing me with the time and financial means to pursue this project over several years. In particular, I want to give credit to the late Richard Greaves for his encouragement of this project. Also, the staffs at the British National Archives, the Tyne and Wear Archive in Newcastle, the Vienna Kriegsarchiv, the Krupp Archive-Essen, the Skoda Archive-Plzen, the Schneider Archives in Le Creusot, and the Military Historical Archive in Moscow deserve special mention. Portions of the discussion of the Ottoman arms trade first appeared in an article in *The Journal of Military History,* and I would like to express my gratitude to the Military History Society for letting me retain the copyright. Among my colleagues at Florida State, Max Friedman in history and Mark Souva in political science provided a critical reading of the manuscript and made invaluable suggestions for its improvement. Most important of all, I wish to express my gratitude to my wife, Lynn, for her patience and loving support through the long research trips abroad and the countless hours spent writing this book.

Rulers, Guns, and Money

Introduction

Perhaps no questions in history are more directly comparative in nature than those surrounding defense issues. Military matters inherently involve comparisons as any given state's strength or weakness is set within the context of its potential friends and foes. It is not possible to understand the development of global war industries without considering the broader context of the emerging Western, and primarily European, arms production and export market. War industries did not grow up in a vacuum, but rather as recipients of the waves of technological diffusion of new armaments and methods of production emanating from the industrial centers of west and central Europe. For their part, the major arms exporters in Britain, France, Germany, and Austria engineered the weapons and shaped the arms market in which other countries participated.

Beyond the borders of western Europe, the Ottoman Empire, Russia, Japan, and China grappled with the challenges of modernizing their defense sectors. Alongside these large countries, smaller ones in the Balkans and South America similarly confronted the problem of importing military technology from the West with varying degrees of success. By 1914 Russia and Japan had accomplished the most in terms of developing a domestic military industrial base, yet even they had not achieved self-sufficiency in armaments or warship production. For all these modernizing countries the challenges posed by the high costs of foreign expertise, imported materials, and skilled labor placed severe economic strains on state resources. In order to arm as quickly as possible with the most modern armaments, these powers turned to imports to a greater or lesser degree.

As the arms trade expanded it did on occasion transform from arms trade to arms race, and then in turn from arms race to war. When and why did the arms trade change from defensive to offensive? These questions are not new. By and large, the existing historical literature on armaments concentrates exclusively on the decade before 1914 with the greatest attention devoted to major studies of the Great Powers (England, France, Russia, Austria, and Prussia) in the prewar decade and issues of naval rearmament.[1] Amid much quantification social scientists examining the contemporary world since 1945 have generally offered versions of two complementary models to analyze arms races. The action-reaction model emphasizes competitive relations between states and external factors. The domestic structure model looks to internal factors, such as the organization and political structure of a given state. Both approaches treat arms races as political phenomena and adopt a state-centered approach. These studies do not attempt to explain specific motives, although they may offer generalized explanations based on assumptions of rational actor decision-making regarding state security or nonrational factors, such as culture, ideology, and religion.[2] Much of the literature avoids delving into the qualitative aspects of the participants' perceptions, motives, and timing in favor of quantitative assessments of such things as the economics of defense burden in relation to gross national product (GNP). Examination of the qualitative dimensions is extremely difficult due to government and business secrecy that surrounds the process. Lack of access to the inner workings of state officials and defense firms inhibits the research for the contemporary world. Consequently, the decision-makers' actual motivations and processes behind the specific armaments acquisitions have remained obscured in an impenetrable black box.

This work seeks to penetrate that black box. In addressing these issues, it foregoes regression equations and instead concentrates on the qualitative side that deals with participants' perceptions, misperceptions, and motivations. Why did participants import and export arms as they did, and what did it mean for them? In taking up the problem, archival sources serve as the empirical foundation for this study. A historical approach offers a unique advantage to the study of the arms trade, because this archival material of the private suppliers—the European Great Powers' intelligence reports and diplomatic correspondence with and from buyer countries—is no longer politically and personally sensitive to those who were involved. This fact allows for a thorough investigation of the business and state actors in a way that is impossible for the contemporary world. In particular, the historical method helps to clarify the transition from defensive to offensive arms acquisition and if and when policy-makers had grander ambitions than self-defense. The holdings for the Armstrong Company in Newcastle-upon-Tyne, Skoda in Plzen, Krupp in Essen, and Schneider et Cie in Le Creusot have made possible an internal examination of the arms trade from the business side. No arms deal remained secret for long. The military attachés proved very adept at uncovering and verifying information. They compared notes with their colleagues in other embassies, had the ears of company representatives, and often knew the intimate details of contracts and negotiations. The British intelligence reports from the Public Record Office (Kew) consistently proved the most insightful and accurate. For Eastern Europe this work also uses the Austrian reports from the Kriegsarchiv and the Hof-, Haus, und Staatsarchiv in Vienna, the Russian military intelligence reports housed in the Military-Historical Archive in Moscow, and captured German documents available on microfilm. For an understanding of the various buyer states, I have relied on the scholarship of historians with expertise in the specific regions and countries.

This study probes the global arms trade and the world's first significant arms races in the industrial age. It considers the connections among private business, diplomacy, the domestic politics of procurement, and military technology transfers. Around the world many countries faced similar developmental challenges linked to their later industrialization and their emphasis on meeting defense needs in competition with one another. Thus, the world witnessed the birth of the first major military-industrial competitions beginning in the nineteenth century.

The story of the global arms trade consists of multiple narratives; to tell it requires a synthesis of the business side, the Great Power side, and the buyer-country side. Some coordination between the diplomatic relations of the home state and customer country was possible since the supplier firms resided in the core Great Powers in Europe. However, the arms trade story did not simply reflect the diplomatic story. Industrial and banking interests often found themselves at odds with diplomatic or military-strategic interest, with the results showing that generally the history of the arms trade did not conform neatly to the diplomatic history of the period.

For private armaments producers in the nineteenth century, tsarist Russia, the Ottoman Empire, and the Balkan states served as vital markets for rifles and artillery. As an engine for growth, contracts from these East European customers proved crucial for the sustainability of the suppliers from west and central Europe, and overall the region became the most important defense market in the world in the period 1860–1914. While some scholars might choose to classify the Ottoman Empire as Middle Eastern, I have included it in Eastern Europe, because the strategic heart of the empire resided in southeastern Europe between the Balkans and Constantinople; the bulk of the armed forces were deployed there; and its chief challengers also lay in Eastern Europe. The importance of purchases from such states as Russia, the Ottoman Empire, Romania, Bulgaria, Serbia, and Greece lay not just in the vast quantities of war materials ordered, but also in the timing of those orders. Often private firms from the United States, Germany, Austria, and France found their salvation in orders from or business ventures with Eastern European customers at pivotal periods when insignificant domestic sales threatened the companies with financial loss or even collapse. At other times the firms, including British companies, looked east as part of a global export strategy for expansion. For naval sales, East Asia and South America served a similar function. On the receiving end, the buyer states acquired the latest models of military and naval systems at the same time or even slightly ahead of their deployment in the more advanced industrial powers. Since most of these countries completely lacked a domestic military-industrial base, their procurement of state-of-the-art weapons proved integral to the buyers' defense postures.

Private enterprise stood at the center of the nineteenth-century international arms trade. Western artillery firms originally moved toward defense production and away from rails and simple metallurgy because they could not compete with domestic rivals who paid less for raw materials and transport. These future military-industrial firms sought to offset these disadvantages by moving into more specialized engineering. At roughly the same time, Armstrong and Krupp pioneered the breech-loading artillery piece. The German firm Krupp had already moved out of rail and general steel production and into armaments in the early 1860s. The British gun maker Armstrong had started out specializing in hydraulic machinery in 1847 but founded the Elswick Ordnance Works in 1859. After initially importing technology from Britain, private French firms also oriented toward arms exports. Whereas France lacked any private gun producers until 1870, at that time, with government aid, the firms of Schneider-Creusot and St. Chamond adopted the best British machinery. By 1875 these young French firms surpassed the English in commercial steel, and the following year French industries were making tubes for guns, steel shells, and steel armor. Schneider increased its commitments to armaments, and the Loire region emerged as the military-industrial heart of France. In 1886 Schneider gave up rail production entirely and switched over to the business of war supplies. Similarly, in 1888 the English firm Vickers turned to the manufacture of armaments as its rail and general steel trade fell away, and produced its first artillery piece in 1890. Skoda, situated in Plzen, Czech Republic, made its first strategic moves into the defense business in the 1890s after the Austrian navy accepted the Skoda pattern for quick-fire artillery and the company supplied small caliber quick-fire guns to the army. Krupp, Armstrong, Schneider, and Vickers had all moved into naval construction as well by 1900. The historiography for private European armaments firms covers the period 1854–1914 in the treatment of a number of individual companies. However, business historians have not paid special attention to, or given any detailed appraisals of, the regional aspects of the international arms market, although Western armaments firms operating in Russia, such as Vickers and Schneider, have received some mention.[3]

Although state arsenals supplied much of the armaments domestically for home governments of the European Great Powers, private firms overwhelmingly dominated the arms export markets. For this reason, armaments firms found exports essential for their viability. Home governments simply did not purchase enough to sustain the private firms. States in Eastern Europe, East Asia, and South America all purchased state-of-the-art weapons systems directly from the firms. Among these regions, Eastern Europe functioned as the most important. During the period 1854–1914 the governments of the Ottoman Empire, Russia, and the newly independent Balkan states (Greece, Serbia, Bulgaria, Romania) proved the most consistent long-term customers. Indeed, in the early days of the various armaments firms, often orders from these governments made the difference between business survival and bankruptcy. Toward the turn of the century East Asia and, to a lesser extent, South America gained in importance, especially for the naval trade. As waves of equipment modernization made older systems obsolete, the surplus weapons became profitable goods in the secondhand gun trade as private merchants acquired government castoffs and retailed them. In this case, Africa absorbed the bulk of the rifles, with Ethiopia leading the way. As the number of private producers increased, business competition grew more intense. By the eve of World War I many of the European private firms had established branch plants or invested directly in joint ventures in foreign countries as the best means to retain market share.

This book elucidates the existing discussions of the arms race leading up to World War I by providing broader scale and scope. In considering more than half a century of case studies rather than a single decade, this longitudinal study allows for a more accurate appraisal of the links between armaments and the outbreaks of war. In addition, a comparison of the regions of Eastern Europe, East Africa, South America, and East Asia significantly broadens the spatial domain by illuminating the global processes beyond western Europe. The results therefore seek to strike a constructive balance between the specifics of each regional case and generalizations for the world as a whole. The goal of this study is to disentangle the threads of diplomatic, military, political, and economic factors in explaining specific outcomes for each country.

In addition, the arms trade had ramifications for imperialism. Scholarly explanations of imperialism have tended to emphasize the disposition of the imperial home countries in terms of their means, opportunity, and motive.[4] The means derived from the technological advantages spawned by industrialization, such as rapid-fire weaponry and steel warships, dramatically bolstered the military capabilities of the imperialist states over the nonindustrial polities. Thus, a huge power imbalance developed that decisively tipped the scales in favor of imperialism. Having acquired overpowering military means, the imperializing powers needed only an opportunity and a reason, whether economic, strategic, or military.

The arms trade connects to the problem of imperialism in two ways. First, the armaments business entangled the threads of economic, strategic, and military interests. By selling armaments to the periphery, firms pursued their own economic interests while convincing their home governments that such sales represented national prestige and could be used as a tool for exercising controlling influence in foreign countries. However, in the process these sales could alter the regional strategic balance with repercussions for great powers and little powers alike. Second, the commercial transfer of arms mitigated the military power differential between industrialized and nonindustrialized states. In effect, the arms trade provided the nonindustrial states with the means either to resist imperialism or to pursue their own.

Around the globe imperialist mentalities informed the perceptions and misperceptions of the arms trade. Diplomats of the Great Powers and their governments believed that they could have influence through the arms supply. As a function of their imperialist mentality they viewed weapon sales as a zero-sum game, where the buyer's acceptance automatically affiliated that state with some supplier power. This bred a blatant condescension that they could purchase the allegiance or support of buyer states by dazzling them with the trinkets and baubles of the arms trade. They acted in accordance with that erroneous belief with embarrassing or sometimes fatal results. Ethiopia provided the most glaring example. The Italians assumed that John Howi Menilek, king of Shewa and then emperor of Ethiopia, could not be an independent operator, but must serve some European side. They therefore plied him with weapons, only to find themselves on the receiving end of those rifles.

In the closing decades of the nineteenth century, imperialism acted as a cause and an effect on the arms trade. The scramble for colonies and spheres of influence in Asia and Africa by European powers provided the external stimulus for others to seek armaments for themselves, either to join in and carve up their own piece of the action (Japan) or to stave off the imperialist threat (Ethiopia). The stakes were highest in Asia since the vulnerable Chinese empire offered a tantalizing prize for many competitors. As the Japanese set their sights on Korea, the Russians frustrated them while angling for Manchuria and more. The Russian shift in focus away from the Balkans and toward East Asia had momentous consequences for the global arms trade. As we shall see, Eastern Europe dropped and East Asia shot up steeply in importance as a regional armaments market precisely because of Russia's new orientation. In East Asia, the Japanese realized that to contend with the Russian behemoth they needed to buy as much as they possibly could. Consequently, Japan became the biggest importer of warships and artillery in the world and entered the ranks of the imperial Great Powers after its victory over Russia in 1905. In Africa it was not accidental that the only country not taken over by the Europeans, Ethiopia, was also the one purposefully connected to the arms trade.

Massive arms imports raised the possibility for the military suppliers to exert extraordinary influence in creating dependent client states, but that outcome did not occur. Arms sales held vital importance for the companies, but the same did not hold true for their home governments. The preponderance of private enterprise in the global arms trade meant that the relationship with the buyer states more closely resembled a business-customer relationship than a patron-client one. The differences between the two types helped to explain why arm sales had such little effectiveness as a tool of diplomacy by the supplier government. Influence is the coin of the political realm, but we do not usually talk about businesses having influence with their customers beyond inducing their customers to buy more. A consumer country's purchase from a firm of a given power did not automatically confer positive political influence by the supplier country's government or even long-term gratitude toward that country. The refusal on the part of a consumer country to buy from a firm of a given country, however, could signify a negative attitude.

To determine the effectiveness of foreign influence through the arms trade, it is first necessary to have a practical definition of the concept of influence. In the context of the arms trade, influence could be seen as the capability of the supplier government to alter the policy of the buyer government in conformity with the desired policy goals of the supplier. This study takes a subtler view of influence. It recognizes that attempts to exercise influence could flow both ways rather than only unidirectionally from supplier state to buyer state. A supplier government or a buyer state might seek to exert influence through the arms trade to achieve other ends, such as lining up an ally or cultivating a protector. Likewise, buyer and supplier governments could try to exert influence commercially in favor of the arms trade through diplomatic channels. Additionally, the buyer country could work to counter the effects of supplier influence by trying to exert some influence of its own. This "counterinfluence" could be exercised by employing multiple suppliers to lessen the potential influence of any individual supplier. In this way, different suppliers had to compete against one another. Influence also took the form of acculturated officer elites in the importer armies and navies, whose members received training abroad and were wooed by foreign firms. The question arises: Did the arms trade bring diplomatic influence, or did diplomatic influence bring the arms trade?

Within the importer states the arms procurement process resided at the intersection of business, politics, and foreign policy. Here the regime type played a determining role in how the arms trade interacted with the buyer country. Autocratic states procured differently than democratic states, which required some kind of legislative approval. Variations existed among autocratic states, depending on whether the autocrat personally intervened in the procurement process (Ottoman Empire, Ethiopia), relied on bureaucratic specialists in the war and naval ministries (Russia), or allowed the decision-making to devolve to provincial bureaucrats (China). For democratic states (either congressional republics or constitutional monarchies), legislative control over appropriations introduced broader political involvement into procurement. The outcomes of budgetary votes, the determinations by formal technical commissions and weapons tests, and the norms of contract bidding processes reveal more than the arms trade and procurement process. They also serve as a gauge of a state's democratization and level of professionalism in the armed forces.

Moreover, the manner of the arms acquisitions helps us assess the role played by influence and corruption. We can conceive of a qualitative spectrum for the procurement process with merit and corruption at opposite ends. If weapons were chosen based on the technical superiority of the weapon as professionally determined based on formal field tests and the formal procedures of contract bidding were duly followed, then merit has won out. If political, business, or diplomatic pressures on the system yielded results contrary to merit, however, then corruption and bribery took precedence. Issues of pricing could occupy an intermediate position in that a lower cost could be a measure of merit, but it could be manipulated by corruption. At times improprieties could and did result in lower prices. Also, financing as a factor could be determined by fair merits, corruption, or imperialist pressures. Were buyer countries forced to take loans, or were they able to shop around for the best terms commercially? As we shall see, cabinets in the buyer countries would fall and the popular press would roil over scandals real and imagined relating to questions of foreign arms contracts and the foreign loans to pay for them. For the citizens of buyer countries financing and taxation had an inverse relationship. The more a government extracted domestic taxes directly, the less it depended on foreign loans. Furthermore, loan terms could intervene into domestic tax collection and divert it from the local treasury directly to foreign financial houses.

We can discern a fairly clear periodization for the arms trade globally, largely determined by technological innovations. During the years 1860–1880, U.S. rifle makers ruled the international market. Breech-loading rifle systems evolved rapidly between 1860 and 1871, with bolt action increasing the rate of fire by employing a breech block or bolt for insertion of the ball and cartridge. During the American Civil War metallic cartridges proved their worth, and after the war U.S. manufacturers held center stage based on the scale and rapidity of their production of breech-loading rifles. Remington, Winchester, and Peabody Tool all gained worldwide notoriety and sales. Meanwhile, the chief artillery suppliers consisted of the British firm Armstrong and the German firm Krupp, the leaders in breech-loading artillery.

In the subsequent era, 1880–1903, the German (Mauser) and Austrian (Steyr) manufacturers held sway for military rifles, and Krupp claimed the crown as the foremost artillery supplier in the world. Österreichische Waffenfabriks-Aktiengesellschaft in Steyr emerged from a rifleworks established by Joseph Werndl in 1864. The firm earned its success based on the Mannlicher system of magazine-fed rifles. Similarly, the Mauser company in Oberndorf rose to prominence thanks to its magazine-fed rifle. The participation of banking interests added a new dimension to the armaments business in this period. Nonpayment by the Ottoman government helped bring about the demise of the U.S. rifle firms, and the desire to avoid bankruptcy provided incentive for the other arms manufacturers to induce banking partners to furnish loans to their customers. By introducing European financial houses into the bidding and contract negotiations, the armaments firms not only guaranteed themselves payment through the issuance of loans to the buyer states, but they also made the financing terms part of the package. Interest rates and payment conditions combined with pricing determined the attractiveness of a bid and the overall costs of armaments purchases. In this process the initiative lay with the industrial concerns. The armaments manufacturers led the financial interests, not vice versa.

Finally, the decade 1904–1914 witnessed the formation of domestic and international business alliances to reduce competition and coordinate market sharing. German and Austrian rifle makers created a cartel and divvied up the Balkan market into agreed-upon spheres of dominance. Simultaneously the artillery competition heated up to unprecedented levels fueled by the arrival of Schneider-Creusot and Skoda.

Procurement of quick-firing artillery became a priority for all armies at the beginning of the twentieth century. French officers engineered the technological breakthrough that opened up the quick-fire era in 1896 by designing a field gun that remained stationary during firing while the barrel itself moved backward. A piston cylinder mounted parallel to the barrel worked as a braking mechanism by absorbing the recoil shock and redirecting it to compress fluid. The fluid in turn reexpanded the brake and pushed the barrel back into its initial firing position. Prior to this innovation, the rate of fire for artillery had reached its limit, and even the use of partial recoil suppressors, such as springs in the mountings or a shovel blade to dig the piece into the ground, could not significantly compensate for the disturbance created by the recoil kick. The new technology dramatically improved the deadliness of artillery by quadrupling the rate of fire and improving direct-fire capabilities, because quick-fire guns did not have to be repositioned after every shot. As a result, acquiring artillery with quick-fire mechanisms dominated army weapons budgets between 1900 and 1914.[5] The French wrested key sales from Krupp's control in the Balkans and posed a serious challenge to the firm in South America. At the same time, Schneider entered into a business alliance with Armstrong and Vickers to break the German hold on the Turks.

The naval trade most consistently had a British face throughout the period 1860–1914, although continental firms made some noteworthy gains. For the naval trade in ironclads, British firms dominated, followed by French, in the era 1860–1880. In the subsequent period, 1880–1903, British shipyards continued their leading position in sales of larger warships, to be joined by German and French firms in the torpedo boat market. During the late 1880s the British developed torpedo catchers, which were replaced by the torpedo-boat destroyer in 1893. By the late 1890s a true destroyer had come of age as an escort vessel. In this same period cruisers had developed from fast vessels to prey on commerce into armored "protected cruisers."[6] Armstrong increasingly gave preference to warship construction in the 1880s. The firm's first three vessels launched from Elswick in 1885 were destined for Eastern Europe: two torpedo cruisers were built for Austria-Hungary, and a cruiser was purchased by the Greek government. Then, in 1887 Romania bought a cruiser. Armstrong, having established itself as a warship builder, did not find good prospects for major contracts in the Balkans in the 1890s, so the company diverted its attention to more promising clients in Latin America and Japan.[7]

The company Schichau in Elbing, Germany, arrived as a leading torpedo-boat producer in the 1880s. Based on the success of its torpedo boat the *Adler*, which at the time was the fastest ship in the world, Schichau beat English and French competitors and sold seventy boats to foreign navies between 1884 and 1890. Eastern Europe again provided the first customers for the new boat in 1884, when both Russia and the Ottoman Empire made purchases. They were followed by Austria-Hungary, Italy, and China (1885), Brazil (1888), and Japan (1890).[8]

French firms moved into the forefront of naval technology by the mid-1880s. In 1872 French firms had taken up the problem of piercing armor instead of smashing it, and by 1874 French naval artillery included high-explosive, armor-piercing steel shells. These technical advancements, along with the manufacture of solid steel armor, proved superior to English compound armor. As of 1884 France had two large private gun factories, Schneider-Creusot and Forges et Chantiers de la Méditerranée, overseen by Gustav Canet, a French engineer who had been trained at Armstrong. Additionally, St. Chamond had a gun plant, and three other smaller ordnance firms existed as well. The French navy, though, continued to supply itself with guns from state works at Ruelle. Denied the major domestic customer, French private firms were forced to look for foreign orders. In 1886, with financial support from Parisian banks, Creusot arranged a syndicate that included the shipyards of La Seyne, Loire, and Gironde to cooperate in the export business. These firms continued to compete domestically, but formed a united front for foreign orders. The only French warship builder not participating in the syndicate was the torpedo-boat enterprise of Jacques-Augustin Normand.[9] The stunning success of French business in naval exports based on superior technology was reflected in a temporary ascendancy over British producers. On the basis of overseas tonnage, French firms more than doubled British sales from 1887 to 1890.

The French technological lead evaporated with the rapid rate of naval innovations. Armor plate advanced when American H. A. Harvey developed nickel steel armor plate in 1890, but in 1894 Krupp developed the superior cemented armor by adding chromium and manganese to nickel steel. German shipyards kept pace roughly with the French. By 1899 Germany had twenty-six shipyards, including eleven private firms on the Baltic Sea, twelve more on the North Sea, and three imperial yards. Krupp-Germaniawerft, Howaldstwerke, Schichau, A. G. Weser, Blohm und Voss, and Vulcan comprised the six most important private yards.[10]

As naval races developed around the Pacific Rim, British producers sold the lion's share of vessels in South America and East Asia. During the 1890s Armstrong scored its most significant success in its sales to Japan. Beginning with a relatively small contract for one torpedo-boat destroyer in 1894, the firm would ultimately also sell the Japanese five cruisers and two battleships. If Japan proved Armstrong's single best customer, South America held the position of most important region. Argentina, Brazil, and Chile bought three cruisers each. Additionally, Argentina bought one destroyer, Brazil one gunboat, and Chile one training vessel.[11]

Like the land armaments trade, the naval business, too, gained momentum between 1906 and 1914 as dreadnought mania seized large and small powers alike. The British revolutionized capital ship design with the adoption of the new "superbattleship" HMS *Dreadnought* in 1906. Seemingly overnight this warship rendered all others obsolete as it proved faster, heavier, and more heavily armed than any other. Compared to previous battleships, which carried two twelve-inch guns, the *Dreadnought* possessed ten such guns. In addition, the ship was the first to have steam turbine engines.[12] Almost immediately all naval powers and naval power aspirants clamored for the revolutionary warship. Alongside the famous Anglo-German naval race of the period, lesser powers—such as Brazil, Argentina, Chile, the Ottoman Empire, and Greece—joined the fray by bidding for battleships with ever-increasing tonnage.

The fierce naval competition led to the establishment of new construction facilities and branch plants provided by European firms within the buyer countries. The British maintained their preeminence, and Vickers in partnership with Armstrong took pride of place in the naval business. In this manner the process of transformation of the arms trade from discrete regional components into an interactive global system reached its culmination. Just as the world economy itself had reached an apogee of integration by 1914, so, too, the arms trade became an interconnected web. The armaments firms established filial plants abroad, and production was linked through subcontracting, patents, and funding by multinational financing and stock investment.

Firms developed their own kind of informal imperialism. The big armaments firms formed alliances and carved out spheres of influence, with or without the knowledge and approval of their home governments. Like the political version, informal imperialism in the business world meant finding local collaborative elites. The firms cultivated native officers in the armed forces and offered them, off the record, a share in the armaments contracts through the payment of commissions or other financial blandishments. Also, the banks and the industrial firms called for "compensation" in their negotiations for contracts, thus invoking the language of Great Power diplomacy. The analogy should not be pushed too far, however, because ultimately the arms trade provided the buyer with the means to resist. Gunboat diplomacy did not work when the imperialist was trying to sell gunboats.

1

Arsenals of Autocracy

Across the globe, the period from the 1860s to the early 1880s witnessed attempts by various states to establish a domestic base to manufacture weapons. Throughout the world governments tended to follow a well-established tradition of relying on state arsenals for armaments needs. Consequently, private manufacturers could not count on their own states to provide enough sales to stay in business, nor did they receive significant diplomatic support when the firms turned to exporting war products as an alternative. Arms exports figured so prominently in the business of private firms because foreign sales offered the only way to preserve plants and to research and develop expertise.

U.S. rifle suppliers experienced a dramatic rise and fall between the end of the American Civil War and 1879. In the process, U.S.-made arms spread to Europe, East Asia, Latin America, and the Middle East. Overall, the Ottoman Empire became the single biggest customer for American weapons. Faced with tremendous overcapacity and a rapidly evaporating domestic market for military arms, U.S. firms eagerly sought foreign sales after 1865. In the years 1865–1870, American suppliers, including the U.S. government, exported approximately 1.5 million rifles. Among the top-five U.S. small arms firms (Remington, Colt, Winchester, Smith and Wesson, and Providence Tool Company) Remington was the largest and had by itself exported more than 500,000 arms.[1] In the 1870s, however, Providence Tool surged ahead with even greater sales to the Ottoman Empire.

Undoubtedly, quick and dramatic changes in technology played a driving role in creating the world firearms market between 1856 and 1878.

With the American development of firearms production based on interchangeable parts and machine-tool precision, European states followed suit. During the 1860s the manufacture of breech-loading rifles using steel barrels superseded muzzle-loaded smoothbores possessing iron barrels. Initially, governments sought to convert old smoothbores into breechloaders. For example, Britain converted old Enfield muzzleloaders into breech-loading Sniders using the system invented by the American Jacob Snider. Other countries, desiring to save money, tried to transform old materials. Interim solutions involved adding breechblock mechanisms to transform muzzleloaders into breechloaders. In addition, metallic cartridges replaced conical bullets. Eventually, the conversion models went by the wayside and governments adopted true breechloaders. As an example, in 1871 Britain officially adopted the Martini-Henry breechloader and the Enfield factory began its production.[2]

For Russia and the Ottomans the arms trade at home signified the restoration of their military capabilities after their poor showing in the Crimean War. The two states had different motivations behind their respective armaments modernizations, however. The Crimean defeat called into question Russia's status as a Great Power. Therefore, St. Petersburg considered modernizing its domestic military and naval production capability. The Ottomans had a vision that was less grandiose for their armaments modernization. Having fallen out of the Great Power ranks in the previous century, the Turks devoted most of their efforts to improving their armed forces' equipment as a means to impress the Russians and possibly deter them. The Ottoman government did not commit itself only to the costly goal of domestic self-sufficiency in armaments. Instead, the Turks found it perfectly acceptable to rely on arms imports of the latest models. Russia had greater wealth and primarily pursued domestic production. Although Russia imported vast quantities, in proportion to the size of Russian armed forces these acquisitions barely made a dent. The Ottomans became dependent on imports, but did not really suffer a qualitative or quantitative disadvantage versus Russia.

Importing the new weapons was undoubtedly the easiest method to acquire new weaponry because it avoided the high expenditures of time and money necessary to establish the new arms industries. But it could lead to absolute dependence on foreign suppliers. Such a position was unacceptable to Russia, and so to gain the technology itself, that country contracted with a foreign firm for a certain number of weapons, and then obtained all the plant machinery on completion of the order. This approach gave Russia domestic control over arms production while providing a way to familiarize Russian workers with the newer equipment.

The successful arsenal modernization of Russia and the growing import dependence of the Ottomans found a parallel in China and Japan. Both states recognized the need to possess Western-style armaments, and both governments embarked on arsenal modernization. Yet China experienced financial and political difficulties and produced disappointing results comparable to the Ottomans, whereas Japan began establishing modern domestic facilities using strategies analogous to what Russia had done.

As an arms buyer and as an imperialist actor, Russia helped to stimulate arms imports across Eurasia. Russian rivalry with the Ottomans turned both states into huge customers between the Crimean War and the Russo-Turkish war of 1877–1878. Simultaneously, Russian expansion across Central Asia in the 1860s and early 1870s posed a challenge to China in Chinese Turkestan. Faced with the prospects of a Russian military threat on China's frontier and a developing Japanese naval threat to the nation's coast, Chinese officials ultimately assessed the land threat as the graver in 1874, and Chinese arms imports increased and undermined the domestic building program.

During the nineteenth century the primary cause of the Ottoman war industries' decline was financial. Government revenues were the lifeblood of the war industries, and by the nineteenth century the "Sick Man of Europe" clearly had a circulatory problem. In the early 1800s Ottoman public revenue was approximately £2.25 million to £3.75 million, compared to the British average of £16.8 million for 1787–1790. Attempts to raise revenues failed to produce tangible results, and the Ottomans were unable to cover the costs of reforms and continual wars. When they had the money, the Ottomans chose to maintain their armed forces. During the 1830s the army claimed 70 percent of the empire's total revenues.[3] The government simply ran out of money in the 1840s, and consequently the war industries began to atrophy.

State borrowing offered an apparent solution to the problem. Beginning in 1841 the state issued short-term bonds to pay the war industry. In 1860 Ottoman state revenues amounted to about £10.5 million, but expenditures were £11 million. Because of financial limitations the empire's treasury was effectively empty. For all its purchases for the army and navy, the Ottoman government had to resort to "long and uncertain credit, and this alone would cause everything to be purchased at an extravagant price."[4] Soon internal sources were exhausted, and the only alternative was foreign borrowing. The first foreign loans commenced during the Crimean War (1854–1856), when the Turks borrowed from private bankers of their military allies, Britain and France. After the war Sultan Abdulaziz (1861–1876) continued to spend money, mostly for the construction of palaces, and by 1863 the internal debt had reached 32.5 million Turkish pounds (TL), and the foreign debt was TL40 million. Due to inefficient administration and tax collection the Ottoman government regularly ran in the red, and the state covered its deficits by borrowing more at high rates of interest. Under Abdulaziz as much as a third of government income was used to pay debts. Consequently, during the 1870s the salaries for soldiers and bureaucrats were chronically in arrears.[5]

The Ottoman government operated all the domestic factories that manufactured and maintained war materials. The government department responsible for the various activities was the Ministry of Imperial Ordnance (Tophane-i Amire Nezareti), which was independent of the Ministry of War (Bab-i Seraskeri) and was entrusted with the production, repair, and supply of weapons and military equipment. In addition, the ordnance ministry's main duties included guarding the straits and training technical personnel. The ordnance ministry consisted of defense, communications, and supply departments, and the ministry directly administered the Zeytinburnu factory and powder mills in Constantinople and the powder mills in Anatolia.[6]

During the 1870s the Tophane factory had the potential to serve as a center for military-industrial regeneration within the Ottoman Empire. The government acquired new and costly equipment and took advantage of foreign technicians. Nevertheless, production techniques did not improve. A British assessment of the works in 1872 noted that despite the large sums lavished on Tophane, the results were disappointing because "the whole scheme has been begun, as it were at the wrong end. A 'machine shop' has been provided, filled with the most perfect and most expensive machinery for finishing guns of the largest size, but there is little or no steam power for working the machines, and the rolling mill and forges are said to be simply useless for the manufacture of heavy ordnance. Such a forge as would be required to bring the gun factory into full operation would cost more than £100,000, and could not be completed for several years."[7] While the Ottomans had paid for the machinery, they neglected the human resources side of the equation. The private British engineers at Tophane did not receive their pay because, in the words of Ottoman war minister Ali Saib Pasha, "They had not fulfilled their engagement by rendering efficient the workmen they had undertaken to instruct."[8] When the British ambassador spoke to the war ministry concerning the Tophane engineers, Ali Saib promised to examine the claims and to act in an equitable manner. The case dragged on for years, however.[9] Tophane and the naval arsenal (Tersane) both suffered from lack of skilled labor. Conscription provided most of the workers, and after their service was completed, the workers departed. This labor system did not yield good production, but it did allow the government to pay wages irregularly and at levels below civilian rates.[10]

After maintaining a respectable naval construction capability for the first part of the century, the Porte's position deteriorated. Two events combined to spell the end of significant domestic naval production in the 1850s. The first was, of course, the Crimean War. The defeat at Sinop crippled Ottoman naval strength by destroying the existing fleet. To compensate for the loss, the Porte purchased warships from abroad. In 1854, the Ottomans bought a large number of ships for the first time from its wartime allies, Britain and France. In all, eight warships were purchased, paid for with a foreign loan.[11] The second event was the development of steam-powered ironclads. The necessity of maintaining some kind of naval force comparable to the European ironclads resulted in more foreign purchases. As a result, the foreign complement steadily rose within the Ottoman navy. In the years 1859–1868, thirty ships were purchased abroad and thirteen were manufactured domestically. The ships manufactured in Britain and France carried just over half of the navy's guns and most of the tonnage.[12]

The British government demonstrated a standoffish attitude about involvement in the arms business. In 1863 the Ottoman government became interested in ironclads that were being manufactured at Birkenhead. The British government informed the sultan that if he wanted to buy the ships, he would have to deal directly with the ships' current owner, the Laird Company.[13] The British Admiralty denied an Ottoman request to have three Turkish ships fitted with their engines in one of the royal dockyards in England. In 1864, Sir Henry Bulwer, Britain's ambassador in Constantinople, worried greatly about the possibility of French influence gaining control in the Turkish navy. He wrote:

> One of my principal cares has been to keep the Turkish navy as much as possible under the influence of English feeling . . . A great and important policy is at stake, which at no distant time may involve the independence of the Mediterranean; every ship in that sea which is not British or Turkish will infallibly be against England; and I venture to think we should sacrifice any minor considerations, and incur some amount of inconveniences in order to create a close connection between the two services . . . If the Turkish navy is once led to rely on French assistance in French Dockyards, and brought in contact with the French marine, we shall for certain lose that assistance we may now expect from it . . . in what may seem a trifling incident lies a most important interest . . . I hold that every attention we give to the Turkish navy, and to our intimate relations with it is attention well bestowed.[14]

Bulwer fretted that the French had closer and more convenient ports, and if the Ottomans accepted the French offers, then the Turkish navy would ultimately become a tool of the French.[15]

France proved more active in the arms business than was England. The French proposed to back a company with considerable capital to establish docks on a large scale at Constantinople. As reported by Bulwer, the plan involved taking hold of all the shipbuilding on the Bosporus, and then packing the arsenal with French workers. The sultan assured Bulwer that he understood the aim and had instructed the grand vizier that such a scheme should not be approved. The plan was thus abandoned, "but in order to please the French Government and Mssr. De Moustir, a small ironclad was ordered in France for the service of the Danube."[16] La Seyne, a shipyard in France, gained a sale thanks to lobbying by the French government.

Later, the Ottoman government made an effort to produce some of its own ships. In the 1860s, Abdulaziz imported expensive machinery and raw materials to expand his country's naval-building capacity. He sought a Turkish ironclad completed from stem to stern on the Golden Horn, including the boiler, the necessary steam engine, and plate-rolling mills.

Most of the necessary materials and know-how had to be imported, however. The financing of this operation came from Britain and France.[17]

The sultan's motivation for a Turkish ironclad seemed calculated to impress the Russians with Turkish fortitude after Russia resumed its Black Sea naval forces in 1871. At an audience with the sultan in November 1871, the Russian representative was informed that the Ottoman government had just decided to build a very large ironclad. The sultan urged the Russian to visit the naval arsenal and to see the progress that had been made in the eight years since his last tour. The British ambassador, Sir Henry Elliot, observed:

> There was no doubt a touch of malice in the recommendation of the Sultan for the ambassador to go look at the naval establishments, and it is probably that he also meant him to understand that the order for the new big ship was, in part at least, the consequence of the denunciation of the neutralization of the Black Sea. I had still entertained some hope of being able to obtain the abandonment of the needless expenditure of money involved in the building of this ship, but the Sultan having thus announced his intention to the Russian ambassador, all further effort would clearly be vain. The Grand Vizier sees this and deplores it.[18]

In sharp contrast to the Ottomans, the Russian navy in 1877 was composed of ships made in Russia, and the tsarist government could take considerable pride in this accomplishment. At the outbreak of the Crimean War in 1854 the Russian fleet had consisted of 446 war vessels of all sorts, of which 65 were steamships. English-built ships accounted for two-thirds of the steamships. The Baltic Works, a private factory, undertook building large ships, the hulls having been built at Admiralty or Nevskii, and the machinery at Baird's or Kolpino. In 1861 Baltic Works produced the first Russian ironclad gunboat, followed by Russia's first oceangoing armored ship, the *Petr Veliki,* in 1869.[19] In 1877 the Russian Admiralty bought these works. The naval ministry built engines at Izhora and Kronstadt and ordered machinery from Russian, Finnish, and foreign workshops. Imported machinery cost the navy 137.5 rubles per indicated horsepower, while Russian-made engines cost 176.3 rubles. Private Russian works cost only 150.1 rubles per horsepower.[20] Russia opted to build its navy domestically, although it was more expensive to do so. Only one vessel had any foreign connection, an ironclad frigate built with the aid of British engineers in St. Petersburg and launched in September 1866. Altogether, Russia had twenty-four ironclads, one hundred gunboats, and miscellaneous other vessels in 1877.[21] The ships of the Russian navy stacked up well when compared to those of the Ottoman fleet, which were mostly foreign made.

At the end of the 1860s the Ottoman government began a rifle modernization program based on American weapons. In 1869 the U.S. government successfully sold surplus muzzleloaders to the Ottoman government. The Turks bought hundreds of thousands of Enfield rifles from the Springfield Armory after the weapons had been cleaned and repaired. In 1869 Blacque Bey, the Ottoman ambassador in Washington, inquired about purchasing 114,000 rifles.[22] In May 1870 the grand vizier confirmed that the Ottoman government, with great satisfaction, had purchased 200,000 surplus rifles from United States.[23] The Turks followed this with another American weapons purchase for 230 Gatling guns at $950 each.[24]

Competition for large Turkish contracts between private U.S. rifle makers led to some unsavory business practices and demonstrated the susceptibility of Ottoman officials to bribes. Not limiting themselves to surplus weapons, the Ottomans entered into agreements with private U.S. firms to produce new weapons. In 1871 an Ottoman contract with Winchester for 30,000 Model 1866 muskets caused double shifts and overtime at the plant in New Haven, Connecticut.[25] In May 1872 the Turks approached the Providence Tool Company,[26] and during Turkish rifle trials in July, Providence's Peabody rifle won the support of the war ministry's council. One of Providence's sales agents in Constantinople observed, however, that "Remington is spending money right and left to secure a favorable report. Never mind any report made that his system will be accepted because the Grand Council is dead against it."[27] In Providence's assessment only Winchester posed serious competition.[28] To hasten the delivery and thus make the bid more attractive, Providence proposed purchasing the Henry barrels in England as a way to cut costs and to save time. Blacque Bey knew that Providence Tool owned the exclusive license for manufacturing the Martini-Henry in the United States and consequently realized that Winchester could not use the patents.[29]

The announcement that the Ottoman government had granted the contract to Winchester stunned Providence Tool. John Anthony, head of Providence Tool, fired off an angry letter to Winchester in protest, and threatened legal action for patent infringement. For his part, O. F. Winchester denied he was violating any patent and in turn protested what he considered the dishonorable behavior of Providence's agent in Constantinople, an Armenian named Azarian.[30] In a subsequent letter Winchester complained to Anthony:

> I believe you to be deceived and the dupe of rascals, in which case I desire to feel that you are not morally responsible. My object in taking this contract was to bring and secure the business to this country. It is in my power to retain it or not, and to put you in the way of making money out of it if you desire. The only way you can do this is through me, because you will not make one dollar out of me by any patent suits, since I shall not give you a peg on which to hang one.[31]

Winchester's bravado aside, he stood on shaky ground and he knew it. In October 1872 he conceded to Anthony that there was not much he could do to prevent Providence from taking the contract.[32]

In the 1870s Providence Tool achieved predominance by acquiring the massive Ottoman contracts while Winchester fell into a secondary position. Since O. F. Winchester wanted to keep the contract in the United States, he devised a solution. Winchester agreed to assign the contract for 200,000 Martini-Henry guns to Providence or to surrender the contract to the Ottoman government on the condition that the Turks pay the Winchester company for the 22,000 and 30,000 Winchester guns they had contracted previously but had not paid for. Winchester concluded, "This is only reasonable, just and proper, as the failure of your Government to do as they contracted, and the result of the large amount of money due us by them, has rendered me powerless to carry out the contract for the 200,000 Martini-Henry guns, but has and is causing me great distress on my business from debts incurred by us, necessitated in consequence of this failure to pay me promptly."[33] With the urging of the Turkish legation in Washington, Winchester transferred the contract to Providence Tool in January 1873. Providence Tool eventually held multiple contracts for 600,000 rifles,[34] and delivered 280,000 by January 11, 1877.[35] Together, the Providence Tool contracts totaled £1.9 million (approximately $10 million in gold), for which the Turks pledged bonds worth $2.7 million.[36] Winchester did not leave the Turkish business altogether. During the Russo-Turkish war in June 1877 the Ottomans approved another Winchester contract for 30,000 Model 1877 revolvers. Thanks in large measure to Turkish orders, the Winchester Repeating Arms Company earned gross sales of $2,802,564 for 1877 with a net profit of $668,381.[37]

Russian firearm modernization after the Crimean War also relied on American means. Like many other governments in the era, Russia at first converted surplus smoothbores into rifled barrels as an economizing measure. After 1866 Samuel Colt of the United States and the Belgian firm Fallis and Trapman brought rifle conversion to Russian state-owned small arms factories at Tula and Izhevsk. In 1866 the war ministry ordered small arms factories to convert 115,000 muzzleloaders into breechloaders using the percussion system.[38] The barrel conversion served only as a stopgap measure. The Russians intended to adopt an entirely new system, and they contemplated a purchase of 50,000 to 100,000 rifles. They sent Colonel Alexander Gorlov to the Connecticut Valley in New England to investigate American firearms producers in 1867. The first fruits of Gorlov's work came in the form of a business relationship with Colt Firearms. The following year Russia ordered from Colt the Berdan Rifle, model 1, an American-designed rifle that incorporated a variety of Russian modifications. After having adopted the Berdan as its infantry weapon, in October 1869 Russia signed a licensing agreement with Berdan to buy 30,000 from Birmingham Small Arms. These weapons were designated Berdan Number 2.[39] Russia made a similar arrangement with Colt in 1871 regarding 400 Gatling machine guns and a license to manufacture in Russia. After the production run concluded, the Russian government took possession of all the machinery used to manufacture the rifle. Russia installed the equipment in Tula, Izhevsk, and Sestrorets, and domestic production finished the country's rearmament. By 1879 the three Russian factories turned out 1,846,579 Berdan Number 2 rifles.[40]

For its service revolver, Russia again looked to a U.S. firm. Smith and Wesson had established a sales presence in St. Petersburg during the 1860s. In 1871 the Russian government signed a contract with the firm for 20,000 revolvers. This order proved invaluable for the company; it was Smith and Wesson's first significant order for export, and it helped the firm climb out of financial troubles. During the 1870s Smith and Wesson produced $142,333 revolvers for Russia. The American piece continued in Russian service until 1895.[41]

Unlike the rifle trade, the artillery business never had an American phase. From the start European firms dominated, and conflict in Eastern Europe played a direct role in developing two key artillery firms in 1860–1878: Armstrong and Krupp. Armstrong's development as an armaments firm can be traced to the Crimean War. British Lord William Armstrong, hearing about the tremendous hardship experienced by British forces while attempting to move heavy smoothbore guns into position during the Battle of Inkerman in 1854, had engineered the rifled, breechloaded, wrought-iron artillery piece.[42] The new technology dramatically transformed the defense business on land and sea. For Krupp, cannon exports had already reached significant levels in the period 1860–1871. Although domestic sales led exports 3,244 to 2,462 for Krupp's smaller cannon, exports almost doubled domestic sales for heavier caliber pieces.[43] The two biggest artillery customers in the world were the Ottomans and Russia; they were also Krupp's two biggest customers in the years 1854–1880 (see Table 1).

To modernize its artillery, Russia imported from Krupp. Russian orders started with gun-tube trials in 1856 and heavy coastal guns to protect the vulnerable coasts from British or French naval attack. In May 1863 the war ministry ordered muzzle-loading cast steel guns. The Russian contract for 120 tube guns, worth 1.5 million talers, was Krupp's most lucrative and biggest order for cast steel works since the company's inception. As part of the arrangement Colonel N. V. Mayevskii, a Russian engineer and designer, came to Essen and worked with the firm. The Russian-Krupp collaboration yielded successful trials of a new breech-loading gun

Table 1　Krupp artillery exports by region, 1854–1886

Region	1854–1873		1874–1886	
	No.	(%)	No.	(%)
Balkans	1,563	(31.6)	2,179	(25.8)
Russia	1,148	(23.2)	1,948	(23.0)
Asia	239	(4.8)	1,071	(12.7)
Latin America	68	(1.4)	503	(5.9)
Africa	539	(10.9)	48	(0.6)
Total all Krupp artillery	4,950		8,457	

Source: Zdenek Jindra, "Zur Entwicklung und Stellung der kanonenausfuhr der Firma Friedrich Krupp/Essen 1854–1912," *Vierteljahrschrift für Sozial- und Wirtschafts-geschichte,* Beiheft 120 (Stuttgart: Franz Steiner Verlag, 1995), 970, 975.

in 1864–1865.[44] In a letter to Kaiser Wilhelm I, Alfred Krupp expressed his appreciation for the Russian order. According to Krupp, the Russian government, through its artillery and navy, had aided in the efforts to perfect the guns. Moreover, Russian support and confidence in Krupp's product had helped Germany gain proper appreciation for Essen.[45]

Less than a decade later, however, Krupp's appreciation of Russian orders had turned to wariness. Early in 1872 the Russians learned about Krupp's new gun and wanted to buy it. The Russian interest posed a problem for Krupp, because he had hoped to arm Prussia first, but Mayevskii (now a general) had appeared with an order for trial guns of the same efficiency. Krupp agreed to construct for Russia a batch of trial guns with similar specifications to the German guns in caliber and design, but much heavier in type and therefore less advantageous for rapid movement. These guns had been finished since the middle of 1871, but Krupp found pretexts for delaying delivery to Russia.[46] In the summer of 1874 Krupp worried about the Russians. He wrote, "In the future we shall have to be on our guard against them. No more is to be expected from them, but we have the consolation that through the Russian orders we got well into the gun making business. We need not thank them for that, for they were in need of our help, they did not do it for the love of us."[47]

As part of the general strategy to domesticate armaments production, the Russians tried to persuade Krupp to establish a branch plant in Russia in 1869. To foster a long-term relationship, the tsarist government was prepared to grant Krupp terms for ninety-nine years. Krupp proved extremely reluctant to venture into Russia. He raised all kinds of objections to extract maximum concessions from Russia. Fundamentally, Krupp did not want either to risk money or to take capable men away from Essen. He expressed confidence that his firm could always supply Russia more cheaply from Essen than it could by manufacturing in that country. In Krupp's estimation, even terms for 200 years provided him insufficient freedom to dispose of any potential Krupp-related business without any hindrance.[48] After the Russo-Turkish war the Russians approached Krupp once again about setting up a plant in Russia. Krupp fended them off, saying that the "world situation" was not good, but in a letter to the company board Krupp wrote, "Entre nous, the world situation may be what it will, but we have nothing, not a man, nor freedom nor experience, to waste in Russia."[49]

Krupp much preferred the idea of exporting to Russia rather than setting up shop there. He considered the Russian venture too risky for business and political reasons. First, he feared that through a factory in Russia the French company Schneider might learn Krupp's techniques and then "he will be able to compete with us in future in all the ends of the earth."[50] Furthermore, Krupp insisted on a monopoly for orders. More important, though, Krupp doubted that Russia would experience domestic tranquility, and he worried that Russia would adopt a hostile attitude toward Germany. Aware of Russian difficulties with production delays and mounting costs in domestic production in 1880, Alfred Krupp advised Friedrich Alfred Krupp to inquire whether the Russians might favor the firm with an order for 4 million to 5 million rubles. Alfred Krupp knew that Russian officials had expressed dissatisfaction with their domestic works, and he hoped that the unfilled contracts might come to his company.[51]

On the Russian side, War Minister Dmitrii Miliutin did not want to rely on imports from Krupp indefinitely. Equipping the artillery branch with steel field pieces followed the strategy of using foreign expertise to supplement Russian production. Russia's inability to manufacture satisfactory steel guns meant buying from Krupp for the short term. However, in the long term Russia's future artillery required the development of a native steel-casting industry. The Russian government worked strenuously to establish domestic production. At the beginning of the Russo-Turkish war in 1877 Russia possessed only bronze field artillery. Miliutin could point to substantial progress by 1881. By then Russia had bought 2,232 Krupp steel guns, but also produced 2,652 steel guns from plants in Perm and the Obukhov factory in St. Petersburg.[52]

Across the Black Sea, the Turks, too, looked to Krupp to modernize their artillery in the 1860s and 1870s. Krupp's business in Turkey began with an inquiry in 1860 followed by the delivery of a test gun in 1861. Pleased with the product, the Turks ordered 108 pieces to be delivered by 1864. Nevertheless, by 1869 most of the artillery pieces in Turkish service were still old smoothbores. They did possess some rifled artillery, but only 200 rifled guns had been issued to troops. In 1875 the Ottomans placed an order with Krupp for 500 steel guns for TL900,000. Along with guns for the army, the Turks also decided to equip their navy with Krupp guns at this time, replacing Armstrong naval guns.[53]

The new Krupp guns represented a significant improvement in the quality of Turkish armaments, but the German firm did not advance solely on the merits of its product. The grand vizier expressed concern about squandering this money because there was "much corruption in the mode which the contracts were entered into."[54] Talk of corruption could be found on Krupp's side as well. The way in which Krupp operated in Turkey can be gleaned from a letter sent in September 1878, a request from Otto Dingler, Krupp's agent in Constantinople, for additional funds for baksheesh (a tip or bribe) to pay his anonymous friend.[55] Overall, though, the Turks were extremely pleased with the performance of their new German artillery. The Ottomans placed significantly large orders with Krupp in the mid-1870s, and had already provided the new weapons to their troops by the outbreak of the Russo-Turkish war in 1877. In the eastern fortress of Kars, Ottoman artillery consisted of bronze 15cm guns, constructed at Constantinople, and Krupp steel guns. Some of the steel guns had been sent out from Essen finished; others were bored at Constantinople. The bronze guns had the same range and effect as the steel, but after 1,000 rounds the grooves become slightly rounded. During the siege of Kars the maximum number of rounds from a single gun was fired by a steel Krupp (460) while the maximum for the bronze pattern was 160 rounds.[56]

Alfred Krupp, for his part, was ecstatic about the superior results from the Turks, and he used the war to drum up orders from other countries. According to the director general of artillery in Belgium, Krupp showed him a telegram from one of his agents who had been sent to Osman Pasha at Plevna to report on the working of the Krupp guns furnished to the Turkish army. The agent reported that the guns outperformed the Russians in precision and range of their fire. "Mr. Krupp mentioned that he has made 2800 guns for the Turkish army and navy all those lately furnished being of the last improved pattern. The Turkish Artillery is only second to that of Germany and as regards the possession of the guns in question each superior to that of Russia. In the opinion of Mr. Krupp's agent the Turkish victory is chiefly owing to the perfection of these new guns."[57] The sales pitch worked, and the Belgians ordered the new guns.

The domestic threat of the Taiping Rebellion (1850–1864) was the driving force behind China's arsenal modernization, with an emphasis on provincial supply. Because of this, Chinese policies for weapons and technology imports remained very decentralized and uncoordinated. During the Taiping Rebellion, Chinese officials stressed improvement in the provincial armaments supply to suppress the rebels. Chinese provincial authorities found ways to supply Western-style weapons on their own. Tseng Kuo-fan and Tso Tsung-t'ang set up ordnance plants in Hunan. In December 1861 they founded ammunition and powder plants as well as an arsenal at Anking. Li Hung-chang, Tseng's lieutenant, was acting governor in Kiangsu. In 1863 Dr. Halliday Macartney, a British doctor, had impressed on Li the advantage in manufacturing ammunition to avoid the high costs of imports. Li took Macartney's advice and established Remington rifle production at the Kiangnan arsenal in Shanghai. With the aid of four foreign technicians, Kiangnan was able to produce Remington breechloaders with iron barrels, and in 1872 initiated Remington cartridge production. The arsenal's output reached 3,500 rifles per year by 1875.[58] After importing the machinery from Greenwood and Battley of Leeds, England, the provincial arsenal at Tientsin also developed the capacity to manufacture 5,000 Remington pattern rifles a month.[59]

Chinese officials also commenced domestic naval construction. In 1866 Tso Tsung-tang proposed creating the Foochow naval yard to make modern warships using French technical assistance.[60] Under the direction of two Frenchmen, Lieutenant Prosper Giquel and Paul d'Aiguebelle, Foochow would serve as a training facility and for five years European engineers and workmen would teach the Chinese all the necessary theory and practice of vessel construction. Between 1869 and 1874 a staff of seventy Europeans was under contract, and by the end of February 1874 the arsenal had completed fifteen vessels. The engines of earlier vessels were ordered from Europe, but later ones were constructed at the arsenal.[61] Foochow had no docks, but did have three building slips, five steam hammers, and rolling mills. In a memorandum dated December 19, 1872, Prosper Giquel, chief director of the Foochow Arsenal, laid out the objectives of the arsenal: (1) to build steamers (hulls and engines complete); (2) to teach the Chinese to build steamers and navigate them at sea; (3) to manufacture into bars and plates native iron found in China, old foreign iron purchased in China, and imported copper from Japan.[62] The Chinese workers showed themselves to be skillful, but the machinery suffered from neglect. In addition, the cost of the work at Foochow proved more expensive than purchasing similar articles in the open market. For example, the iron plates rolled in the arsenal could be obtained from England at about half the cost of their manufacture. Nevertheless the government had a stake in keeping the yards working.[63]

In the wake of the Taiwan crisis with Japan, a major Chinese debate took place in 1874–1875. The central issue concerned China's priority for defense. Should the authorities emphasize as the primary theater the land frontier of Chinese Turkestan (the Russian threat) or the east coast (Japan)? If the maritime theater received priority, then naval development would need to take precedence. The debate pitted Tso, the advocate for the Turkestan frontier, against Li, who favored the navy and coasts. Ultimately, the frontier defense won out, and the bulk of China's resources went into the military campaigns in Turkestan. From 1875 to 1881 the Turkestan frontier absorbed 77 million taels. In comparison, the annual naval budget amounted to 4 million taels. As author Thomas Kennedy argued, by opting for the frontier strategy China seriously undermined shipbuilding. The campaigns in Turkestan came at the expense of modernization and increased production by coastal arsenals.[64]

The frontier policy led to increasing reliance on imported weapons, but provincial authorities did not adopt any standardization. Along with the Remington rifles, Chinese officials also bought 13,500 German Mauser rifles between 1874 and 1881. The artillery purchases included Armstrong and Krupp pieces, although the German firm was gaining the edge. In 1875 China purchased 403 foreign cannon (mostly Armstrong pieces) and 20,745 rifles through Shanghai alone. Simultaneously 70–80 new Krupp guns arrived at Tientsin. Altogether, from 1874 to 1886 China bought 850 Krupp guns.[65]

Domestic production did not cease but, starved of revenues, the arsenals could barely manage to hold their own, and what they managed to produce became increasingly outdated. At Kiangnan in 1882 the main work was manufacturing twenty eighty-pound Armstrong muzzleloaders, which required more than a year to complete.[66] John Makenzie, superintendent of Armstrong's gun plant in Newcastle, had established production of steel-barreled muzzleloaders at Kiangnan in 1876, but these muzzleloaders were already obsolete compared to contemporary breech-loading artillery. As another indication of creeping obsolescence, by 1881 the Kiangnan Remington rifles had also become outdated.[67]

After 1874 the northern and southern commissioners' naval import strategies challenged the domestic building program. Li led the way in foreign purchases. He wanted six cruisers and some smaller vessels for his fleet. Through Robert Hart, a British member of China's Maritime Customs Inspectorate, Li ordered four Armstrong gunboats in April 1875, spending £112,800. The first pair arrived from England in November; the second pair departed for China in March 1877. In 1877 Li ordered four more Armstrong gunboats through Hart, and two Armstrong cruisers through Giquel's agency. In July 1881 Armstrong completed the two cruisers, *Chao Yung* and *Yang Wei,* for China. Li bought ten ships through Hart by 1887, when he stopped placing ship orders abroad. Southern Commissioner Li Tsung-his bought four Armstrong gunboats as well, but his were made of steel instead of iron. These boats arrived in 1879. A final three gunboats ordered (two for Peiyang, one for Canton) sailed from Britain in May 1881.[68]

Britain expressed reluctance to supply heavier vessels to China. Chinese officials approached German builders, and German naval sales figured prominently from 1881 to 1884. To counter the three large ironclads (3,000–4,000 tons) Japan had recently ordered, Li sought to acquire China's first battleships. The Chinese thought it prudent to follow the example of other nations, and ordered first-class armor-plated ships. The Chinese navy constantly added warships. In 1882 three ironclads were under construction at Stettin.[69] Vulcan launched China's first battleship, the 7,500-ton *Ting Yuen,* in December 1881. Two other warships were issued from the Stettin yard by 1884. Additionally, the firm Howaldt (Kiel) built two steel cruisers in 1883; these ships joined Chinese naval forces after the war with France.[70] The Chinese fleet also included twenty to thirty foreign-built revenue cruisers mounting two or more Armstrong guns, and twenty others built in China.[71]

Instead of a unified Chinese navy, by 1883 naval forces consisted of four separate fleets: Tientsin, Shanghai, Foochow, and Canton. These four naval fleets were nominally under the control of the two commissioners of coast defense: Li Hung-chang and the viceroy of Nanking, Chang Chih-tung. On the eve of the French war in 1884 China had fifty modern ships, and more than half came from domestic sources. The foreign complement consisted of thirteen Armstrong gunboats, two Armstrong cruisers, and two German cruisers.[72]

The rise in imports posed the problem of standardization. Li Hung-chang had foreseen this difficulty, and in 1875 he urged some degree of uniformity, including the adoption of Krupp heavy guns and Snider and

Remington firearms. Nevertheless, almost a decade later Chinese naval forces still lacked standardization in weapon systems. In the battle of Mawei against the French in 1884, the Chinese deployed British muzzle-loaders, Vavasseurs, and Krupps.[73]

The multiplicity of weapons systems along with the tension between imports and domestic production stemmed from personal political rivalries and divisions within the Chinese administrative structure. As John Rawlinson noted in his seminal study of Chinese naval development, time constraints and superiority of foreign weapons and naval systems alone did not explain Li's growing preference for importing war materials. Issues of personal influence also played a role. The Chinese arsenals and dockyards at Tientsin, Shanghai, Nanking, and Foochow depended on the careers of their founders. Li Hung-chang could exercise only limited control in Tso Tsung-t'ang's dockyard. According to Rawlinson, "As Li's tenure of his northern offices lengthened, he lost influence in the south, where most of the plants were. His maneuvers to reorganize the coasts and centralize purchasing were in part his response to this situation. He was encouraged to seek as much independence as possible, beginning with his 'own' Peiyang fleet. The personal element affected the quality of the arms produced in China and complicated the 'buy-build' problem."[74]

British Major Mark Bell also put his finger on the problem of administrative and provincial competition in his 1882 report on China. In Bell's assessment, China had ample resources for an immense army and a superior navy, but disunity arose from the practice of each provincial government equipping, paying, and assigning officers to its own troops, with no central authority. "The extent of the Empire is so vast, and the government so decentralized, that each province cares for itself only, and is callous to the fate of the rest of China, so long as it itself is untouched. A provincial patriotic feeling exists, but no national patriotic feeling; in fact, the men of North, Mid, and South China are people of antagonistic feelings and would readily join in an internecine war."[75]

China lacked a real central budget, and Peking (now Beijing) relied on the provinces to send quotas from collected revenues. For example, the Foochow arsenal was funded by the foreign customs of Fukhien province.[76] Over 80 percent of the 25 million taels invested in the Kiangnan, Tientsin, and Nanking arsenals came from duties on foreign trade through a percentage of customs revenue at certain ports, an amount that fluctuated year to year.[77] Chinese public revenue was estimated at about £25 million in the late 1870s, and the army cost £15 million. To pay for its armed forces, China relied on customs duties, with more burdens placed on exports than imports.[78]

In stark contrast to the Chinese case, the Japanese Meiji government exerted forceful centralization of arsenals and forged a conscious national policy to protect Japan from Western imperialism. Author Richard Samuels writes, "Military technonationalism, with its national commitment to indigenization, nurturance, and diffusion of know-how, was central to Japanese industrial revolution."[79] Japan possessed a network of arsenals prior to the Meiji restoration. During the late Edo period arsenals had taken two forms: They were either artisan workshops contracted to the central government and managed locally or privately, or they were strictly local enterprises in a particular han. These han (local) arsenals served as a base to build local power to challenge the shogunate's central power. During the 1850s the local lord in Satsuma han introduced Western armaments and manufacturing ahead of the central government. The Saga han started artillery production at Tsukiji in 1850 and by 1854 had made almost 300 cannon. In 1864 Japanese at Tsukiji successfully reverse-engineered British artillery pieces and began their domestic manufacture. The Saga han also pioneered Japanese steamship manufacturing using machinery imported from Britain. The Saga han acquired machinery in 1859, and by 1864 had constructed its own steamship. By the time of the Meiji restoration, han arsenals had developed the most advanced weapons production in Japan. In a similar process, but using Dutch technology, the Shogunate established the Nagasaki Iron Works in 1857, and Nagasaki won the laurels for producing Japan's first steam gunship in 1866.[80] During the 1860s the shogunate began to move away from the Dutch and toward the French. In 1864–1865 the French provided the instructors for the Naval Training School at Yokosuka and the technical assistance to establish a navy yard there as well. The Japanese also bought eight Western warships in the mid-1860s.[81]

With the overthrow of the shogunate, the new Meiji state moved to establish a firm grasp on power. Military matters and armaments were considered a core part of that consolidation process. In 1868 the new government established the Ministry of Military Affairs, including an arms office. As its models, in 1870 the restoration government chose the

British for naval matters and the French for the army.[82] In 1872 the Yokosuka shipyard was expanded. By 1877 the formerly local arsenals were placed securely under central control, and the Japanese arms industry became entirely owned and operated by the Meiji state. Under the Meiji, military spending received top priority. By 1877 military spending accounted for nearly two-thirds of central government investment, and in the 1880s it continued to average more than half.[83]

Having officially designated the British Royal Navy as its model for development, the Japanese logically looked to British firms and expertise in crafting their naval acquisitions. In 1873 a British Naval Mission, headed by Lieutenant Commander Archibald Douglas, began work in Japan, and throughout the 1870s a few small ships were constructed domestically according to Western design. The British built and launched the armored, steel-hulled frigate *Fuso* and armored corvettes *Kongo* and *Hiei* for the Japanese navy in 1877. Samuda Brothers launched *Fuso*, Earle's Shipbuilding at Hull launched *Kongo*, and Milford Haven Shipbuilding at Pembroke launched *Hiei*.[84] As soon as the Japanese had trained their own technical staff they dismissed the French naval architects at Yokosuka in 1876. As a mark of growing Japanese confidence in 1876, the government launched from Yokosuka its first naval vessel constructed without foreign assistance.[85]

The Japanese commitment to the domestication of naval technology lay at the heart of the First Naval Expansion Program of 1882. The program allocated 26.67 million yen to acquire forty-eight vessels over eight years (although two vessels were subsequently dropped), and also called for the development of native shipyards and industries linked to building warships. As part of the plan, twenty-two torpedo boats were to be assembled abroad and broken down for shipment to Japan. By 1883 the British firm Yarrow had supplied four first-class torpedo boats for assembly at Yokosuka. Among the remaining twenty-four ships under the 1882 expansion, fourteen (including two cruisers) were built in Japanese yards. The first Japanese-built warship was the *Katsuragi*, manufactured with British help in 1882. Japan followed this achievement by launching its first iron-hulled ship, the *Maya*, in 1885.[86]

As part of a coherent national strategy to protect its autonomy, the Japanese government chose to pay for its military buildup out of its own resources rather than resort to foreign borrowing. To make ends meet, after 1877 the government sharply reduced its expenditures, privatized state industries, and imposed new taxes. Mitsubishi, Kawasaki, and Ishikawajima-Harima Heavy Industries all started as government arsenals but became privatized in the years 1876–1887. In 1876 the state sold the Ishikawajima shipyard to Hirano Tomizo. In 1878 Kawasaki Shozo built his own shipyard in Tokyo, and in 1887 he bought the Hyogo shipyard from the Meiji government. At the same time the founder of Mitsubishi, Iwasaki Yataro, purchased the government's Nagasaki shipyards. Only Yokosuka and Kure continued as state-run enterprises.[87] These policies brought resources into state hands and enhanced the banking system, but also raised the tax burden for small farmers.[88] Japan, in contrast to China, looked to its land tax as the chief source of revenue. In the latter 1870s the land tax provided 74 percent of Japanese state funds, as compared to just 3 percent from customs.[89] According to Marius B. Jansen, "Japanese taxpayers paid a far higher percentage of their income than their counterparts in Europe, where industrial and consumption taxes had long since passed land taxes as the principal source of government income."[90]

Diplomacy generally took a backseat to business efforts in explaining the choice of purchases. U.S. rifle companies solicited sales on their own without lining up official backing, and given how little power the U.S. government wielded in international affairs at this time, a word from Washington would probably have meant very little. Remington, Providence Tool, Winchester, Colt, and Smith and Wesson succeeded due to the merits of their products. U.S.-made rifles played the dominant role in modernization, either through direct sales, as in the case of the Ottomans, or as models of adaptation for domestic manufacturing, as in Russia and China.

Similarly, the Krupp company rose to the top of the world artillery business. Unable to convince Berlin to purchase his cannon during the early years, Krupp found salvation in exports to Eastern Europe. In the naval arena, British firms continued to exercise dominance over most naval sales, even though they lacked aggressive support for exports from the British government. In fact, the British government consistently held itself aloof from the arms business. Despite France's attempts to intervene more directly in Ottoman naval purchases, the French came in second.

Across Eurasia, the period up to about 1880 can be seen as a time of the region's trying to establish the importance of domestically manufactured weapons, especially rifles. Thus, despite the seemingly over-whelming advantages enjoyed by U.S. producers, the heyday of U.S. market dominance proved fleeting. In 1870, forty-six American small arms manufacturers were in business. By the end of the 1880s only Winchester and Colt had survived, due to successful domestic sales and less dependence on exports. Providence Tool filed for bankruptcy in 1885. Remington gained no new foreign orders for rifles after 1879, and that firm followed Providence Tool into bankruptcy in 1887.[91] In large part, the demise of American suppliers was tied to the financial collapse of their most important buyer. The Ottomans had purchased over 1 million U.S.-made rifles and half a billion cartridges from 1869 to 1879. For the Ottomans, arms imports preceded the Russo-Turkish war of 1877–1878. Although the proximate causes of that war were the Slavic rebellions and Turkish repression in 1875–1876, the Turks were prepared to rely on armed strength rather than give in to Great Power diplomatic pressures for administrative reforms. In this way the acquisition of the latest rifles and artillery convinced the Turks that they could take on the Russians, thereby indirectly precipitating the war in 1877. However, the Russo-Turkish war of 1877–1878 drove the Ottoman government to financial ruin, and in turn the Turks defaulted on their payments.[92] In addition, by the 1880s most of the European countries had developed their own government factories, thereby obviating the need to import rifles from U.S. manufacturers. Greece, Spain, Sweden, Denmark, and China had all procured licenses to manufacture Remington rifles domestically. Meanwhile, France now possessed four state-owned small arms factories and one private producer; Russia had three government plants; and Austria had the largest small arms factory in the world.[93] As Felicia Deyrup observed, "Those American arms makers who shipped gun machinery abroad hastened the eventual independence of foreign countries."[94]

2

Hand-Me-Down Guns:
The Balkans and Ethiopia

The foundations of the secondhand arms trade originated in Europe and rested on unloading older models from state arsenals and reselling them through private arms dealers or state officials. For the state suppliers the intent was not profit but influence, as they ruthlessly undersold the private merchants to secure diplomatic gains. The major rifle manufacturing firms were not the main players in the secondhand arms trade. Instead it fell to the Italians, French, and Russians, and the wares included older Remington and Gras rifles.

In Eastern Europe from 1856 to 1878 the arms trade was important because it was inseparable from the "eastern question." Briefly defined, the eastern question stood at the intersection of two sets of problems rooted in the weakening position of the Ottoman Empire. First, for the European Great Powers the question was whether to prop up the Turks for peace, stability, and the hope of internal reform, or to manage the partition of Ottoman lands in a diplomatically acceptable way. Britain sought to support the Ottoman government and to preserve the status quo of the post-Crimean War. Consequently, the British government opposed the transport of arms into the Christian parts of the empire, and was supported by Austria. France and Russia sought a revisionist course, but for different reasons. The French government, under Napoleon III, eagerly probed for nationalist causes in central and Eastern Europe that might serve to apply pressure on Austria. Russia, under Tsar Alexander II, hoped to undo the Crimean War settlement by which Russia had to neutralize its Black Sea fleet and had retreated from its position in the Ottoman Danubian principalities of Wallachia and Moldavia (the future

Romanian state). As part of these objectives, France and Russia helped to supply arms to the Christian princes within the Ottoman Empire.

The second set of problems concerned the Orthodox Christian peoples, especially in the autonomous principalities of Serbia and Wallachia-Moldavia. Their aspiration for full independence from the Ottoman Empire meant preparing for an armed struggle for liberation. In relation to the arms trade, this meant acquiring military means, often through smuggling, and also seeking recognition of sovereign rights to have armaments and established armed forces.

The eastern question spilled over into East Africa because England's support for the Ottoman status quo meant its disapproval of the arms trade across the southern border of Ottoman Egypt into Christian Ethiopia. After the Ottomans lost sovereignty over Egypt, British strategic and commercial interests increased with the opening of the Suez Canal in 1869, and so Britian continued to oppose the Ethiopian arms trade. However, Italy and France, in response to Ethiopian solicitations, stepped forward as suppliers of surplus weapons by 1880.

Ethiopian leaders found their attempts at developing domestic capacity frustrated by the difficulty of sneaking machinery past European officials, and ultimately John Howi Menilek, king of Shewa and then emperor of Ethiopia, preferred to amass weapons more quickly through direct purchase. Menilek and his designated agents were involved personally in procurement and, consequently, as emperor Menilek developed no government office to handle this function. The making of the modern Ethiopian state was directly tied to the secondhand arms trade and Menilek's distribution of it through patronage to local commanders. By means of the arms trade King Menilek built up his strength to challenge Emperor Yohannes. For Ethiopia the arms trade was about much more than defense. The arms trade went to the heart of international recognition, and maintaining it was a way to assert Ethiopia's sovereignty and legitimacy. The geography of trade routes carried political significance, as Ethiopia was subjected to repeated arms blockades by neighbors. First the Ottomans, then the Khedive, and later the European powers sought to cut the supply line. The annexation of Harar established a secure Ethiopian arms flow, but the French port of Djibouti was needed to complete the route.

Diplomatic motives of the Great Powers in combination with the nationalist aspirations shaped the secondhand arms trade in the Balkans.

As of 1877 only Greece had achieved independence in the Balkans. Serbia, Moldavia, and Wallachia remained under the control of the Ottoman sultan, although all three were permitted to have a prince with the approval of his Ottoman suzerain. In 1859 Alexander Cuza first became prince (hospodar) of Moldavia and then Wallachia, combining the two territories. The Russians and French were pleased with these results and supported Cuza's government. The British and the Turks opposed Cuza. In 1861 the sultan granted Cuza the right to govern Moldavia and Wallachia as a personal union eventually to be called Romania, for his lifetime.

From the start of Cuza's reign the expansion and improvement of the armed forces received priority. The Cuza government and its successor turned to arms imports as an immediate way to make the army of the United Principalities of Wallachia and Moldavia credible as a force to safeguard the union and to erode Ottoman dominance. According to Cuza, "The army had to prove Romania really existed."[1] Beginning in February 1859 Cuza dispatched two agents to Paris to solicit French help in modernizing Romanian forces and purchasing weapons. French arms merchants responded favorably. The Alexis Godillot Company sold Romania 10,000 rifles and contracted to provide 40,000 more, including 28,000 that reached Galaţi in 1859 via a Piedmontese transport. The French government under Napoleon III demonstrated its interest by authorizing a formal French military mission to Romania in 1860.[2]

In December 1860, five merchant vessels under the Sardinian flag passed through Constantinople on their way to the Danube. The Ottoman authorities asked to board the vessels, but this request was denied by the Sardinian legation on grounds that goods were in transit and the Porte therefore had no right to prevent them from proceeding. The cargo included four batteries of field pieces, muskets, cartridges, and ammunition. Turkish officials placed an embargo on the ships and stopped three at Sulina. The Ottomans believed the arms were destined either for Ottoman Danubian provinces or Hungary. However, two vessels eluded the Turks and managed to land their military stores on the River Sereth. The Ottoman government instructed Prince Cuza to seize the military articles and hold them.[3] Thus began a contest between the Romanian prince and his Turkish suzerain over the issue of weapons, and this question quickly engaged the Great Powers as well.

The Sardinian government had initiated the arms shipments as part of its plans for Italian unification. Uncertain about what actions Austria might take in Italy, Giuseppe Garibaldi with help from the Sardinian government had planned an armed insurrection in Hungary. The Hungarian uprising would have led to insurrection in some Turkish provinces as well. Subsequently, the Sardinians had second thoughts and decided not to follow through with these plans. Nevertheless, the Sardinian vessels had discharged their combined cargoes, and Prince Cuza originally put the arms in a warehouse. Most of the Sardinian arms, including 2 million cartridges and 2,000 rifles worth over 500,000 francs, originally came from Marseilles and were sent to Sardinian arsenals by Prince Napoleon in the summer of 1859.[4]

The Great Powers agreed that the arms should be taken out of the principalities, but they disagreed over the question of whose authority should be exercised. The Ottomans insisted that the weapons be sent back to Constantinople. Britain concurred. Likewise, Austria thought the Porte should confiscate them. Meanwhile, the French and Russian representatives opposed handing the weapons to the Turks. Cuza refused to give them up to the Porte, and fundamentally, he made no effort to ship out the arms. The British embassy supported the Porte's right to demand delivery of these arms to Ottoman control. Sir Henry Bulwer, the British ambassador to the Ottomans, noted in a report to Earl Russell that an invasion of the principalities would be an act of war against the Porte, and so the Porte was rightly interested in making sure that the principalities did nothing to provoke such an invasion. Moreover, Bulwer noted that the Ottomans could not calmly submit to the arms staying in the principalities or foreign insurgents bent on revolutionizing or attacking a state on the principalities' borders. The Porte would not consent that Prince Cuza should send the arms to Genoa on his own authority, for such action would effectively acknowledge his sovereignty. Prince Cuza declined to hand over the weapons to the Porte, because he thought that by doing so he would be giving up his rights. Ultimately, the Porte and Cuza turned to Bulwer to resolve the matter. Under British direction, all the arms returned to Constantinople.[5]

Bulwer's resolution of the situation did nothing to end arms shipments into the region. Vessels carrying arms continued up the Danube, now with Serbia as the intended recipient. The importation of gunpowder, cannon, arms of war, and military stores had been prohibited by imperial decree throughout Ottoman domains on March 13, 1862. Nevertheless, in October 1862 a Russian steamer towed a Greek barge up the Danube to deliver stores of muskets, balls, and saltpeter to Serbia. The Russian consul at Galaţi had made the contract for the cargo, and because the Serbian principality had the right by treaty to maintain a militia, the Serbian government could justify the importation of arms and materials as necessary. The fact that the arms on the barge were covered with fish and other goods raised suspicions among British and Austrian observers.[6]

The land routes also witnessed the clandestine flow of arms into the region. In late November 1862, hundreds of carts conveying arms crossed the frontier from Russia into Moldavia with Serbia as their ultimate destination. When British representatives asked Cuza about the arms traffic, the prince dismissed the reports as ridiculous, arguing that 500 wagons could not possibly pass into his territory unnoticed. Despite Cuza's professed ignorance of any such arms convoy, evidence mounted. British officials identified large quantities of Russian arms, perhaps 800 cartloads.[7] British suspicions about Cuza's complicity in smuggling arms to Serbia grew. Eventually, the prince acknowledged the existence of the arms. As a British report observed on December 3, 1862, "Prince Cuza having boldly denied all knowledge of the arms from the 23rd ultimo till yesterday, when he did not confess until he had ascertained that some of the cases were actually at the barriers of Bucharest, together with the open assistance given by the authorities to the persons in charge of the arms, sufficiently justifies the inference that the Moldo-Wallachian Government cannot be expected to act with good faith in this matter."[8]

After Cuza had finally admitted that the arms convoy existed, the British moved to exert pressure on him to prevent the delivery to Serbia. The British told Cuza that if he allowed the arms to go through, they would consider him as "an accomplice in hostile aggression on the Porte."[9] Cuza refused to sequester the Russian arms, basing his decision on the grounds of Moldo-Wallachian autonomy. Russell expressed his exasperation at Cuza. "Whatever privileges and powers this term 'autonomy' may signify," Russell wrote to Consul-General Green, "the privilege of introducing and conveying arms for the purpose of aggression against the Porte is not one of them. Even were Prince Cuza an independent Prince . . . Prince Cuza would have no right to connive at, or be accessory to, preparations for war against the Sultan."[10]

The cooperation and coordination of the arms trade between Romania and Serbia were part of each principality's long-term policy to secure independence for itself through military means. Not only had Cuza sanctioned the transit of 63,000 rifles across Romanian territory to Serbia between November and December 1862, but he had also ensured the safe delivery of the convoys by protecting them with Romanian troops from Turkish attack. The Serbian government appreciatively reciprocated by selling twenty-four pieces of field artillery to Romania.[11] According to Ilija Garašanin, Serbia's minister of foreign affairs, "Though prince Cuza agreed only to the transit of arms meant for us through his territory, once we were in danger he proved ready to help us by all means and to keep his promise even at the price of sacrifice . . . Therefore, we are proud to be the allies of a nation whose leader knows how to protect his rights so nobly against foreign aggression."[12]

Prince Michael of Serbia maintained that he could import arms in any quantity he desired, based on Serbia's commercial freedoms as spelled out in the Treaty of Paris. Serbia denied any hostile intent toward the Ottomans, and the prince insisted that he needed the arms to keep public order.[13] In a meeting with the British acting consul on December 28, 1862, Prince Michael expressed his frustration. Raising his voice, he said, "I do not see with what right I can be prevented from getting arms; and yet the Porte will seek with redoubled vigilance to keep back every fresh convoy of arms from the frontier of Serbia. I make no secret of this: that, in addition to the existing powder-mills and cannon foundries, I shall establish a gun factory, by means of which the want of arms for the country will be provided for, for all time. Who will be able to prevent my doing so?"[14]

As Prince Michael had indicated, the Serbian response to the difficulties with arms imports was to develop a domestic manufacturing capacity. Accordingly, the main government arsenal at Kragujevać saw increased activity. The facilities included twenty machines for rifling barrels and also possessed the capability to construct rifles in their entirety. Following French models, the Serbs could make 500 new rifles per month. A British royal engineer, Major Edward Gordon, visited Kragujevać in June 1863 and reported that the Serbs had already rifled almost all the Russian muskets that had recently arrived and converted their flintlocks into the percussion system. Serbia had established the Kragujevać munitions factory in 1850, but as of 1863 only 7,000 rifles were in hand.

Russia had promised Prince Michael 70,000 more, but only 31,000 had been received. These Russian smoothbore muzzleloaders were retrofitted at Kragujevać into breechloaders with rifle bores. By 1866 the arsenal was producing 500 new rifles and 5,000 retrofits a month. In addition to firearms, Kragujevać had been producing muzzle-loading artillery. By 1867 it seemed plausible that Serbia could equip an army of 100,000 from domestic capacity, although one-sixth of the arsenal's workers (250 out of 1,500) were foreigners.[15]

The level of Russian participation in the arms transfers caused friction with Britain, since the two governments disagreed on Serbia's rights and whether the acquisition of these arms constituted a defensive or offensive action. From the Russian perspective, the Serbian government had a right by treaty to arm its soldiers and people, and the Serbs had imported these goods by a vessel using the Danube. On this point Prussia agreed with Russia. Germany's Prince Otto von Bismarck, too, believed that the Serbs were entitled to purchase abroad as many arms as they deemed necessary to maintain order within the principality. Furthermore, the Russians viewed the Porte's right to prohibit the importation of arms and munitions of war into Serbia as a question open to discussion. Given the current climate, enforcing the prohibition would be unwise as it might stir up the Porte's Christian population. Under the circumstances, the Great Powers should refrain from aggravating the Serbian government by preventing its execution of its natural rights.[16] Russell considered Serbia's intent as an offensive posture against the Turks. He wrote to Consul-General Longworth, "NOBODY [sic] is going to attack Servia, and Servia must not attack Turkey. The arms, however, can be intended for no other purpose than to make war against Turkey, and they ought not to be allowed to pass through the Moldo-Wallachian provinces, which are a component part of the Turkish Empire." The British further advocated that the weapons should be seized and sent to Constantinople.[17] The British position went so far as to assert that even if Serbia were fully independent, rather than an autonomous principality, Prince Michael would not have "the right to arm against a neighboring state without proof of aggressive intentions on the part of that state."[18]

The British had good reason for their suspicions toward Serbia, Russia, and the Moldo-Wallachian principalities, given the secrecy, the denials, and the quantities of weapons. Enormous amounts of war materials, out of proportion to the alleged purposes, had secretly crossed the

Russian frontier, and some of the arms were concealed under commercial goods—a precaution inconsistent with a legitimate purpose. Prince Cuza and Nikolai Karlovich Giers had denied their existence until confronted with the evidence as ascertained by the Austrian and British consuls. According to Cuza, the shipments had totaled 40,000 muskets; the Serbs said they had received 20,000. The British also knew that Prince Michael had abstained from declaring war only because his foreign officers deemed Serbia unprepared. It appeared evident that Serbia sought to develop a larger armed force, and such a force could tempt the prince to violence.[19]

The covert dissemination of arms into Serbia through Moldovo-Wallachia started anew in 1863. Once again, Russia served as the source. Russian military authorities used the passage of two infantry battalions of the 13th Division as a cover for the arms shipments from Kherson. While making it appear that the arms were being sent to marching troops, the Russians dispersed small consignments of arms from Nikolaev to some point near the Danube, as soon as the spring thaw cleared the ice on the River Bug. According to a Serbian informer on the staff of engineers, the Russians were providing 50,000–60,000 arms to Serbia, paid for in part by the Russian government and in part by money raised from subscriptions from wealthy patrons in St. Petersburg. By April, Turkish authorities reported that the arms were being shipped from Kherson to the Bessarabian frontier. Russian Foreign Minister Gorchakov informed the British that no further sale or transport of arms had been sanctioned from Russian arsenals. Also, the arms previously forwarded to Serbia had been sold, not donated, by the Russian government.[20]

In addition to the clandestine importation of arms, Cuza's government also pursued native production and importation of weapons through formal, public channels. In February 1860 the government began construction of an arsenal at Malmaisson, and later a gun factory and pyrotechnics works at Dealul Spirei for explosives and artillery on a Belgian model. The oversight of the operations fell to Belgian Major Herck. During August 1862 Cuza had sent Herck, an artillery officer, on a secret mission to Serbia to learn about the foundry at Kragujevać and to ascertain what Serbia would be willing to supply to Romania. In 1865 Cuza sent Herck, now a colonel and manager of the Bucharest Arsenal, to Serbia to give Prince Michael prototypes of arms manufactured in Bucharest. In terms of public procurement, in November 1863 the Romanian parliament voted a credit of 7.98 million lei to buy rifles, cannon, powder, and shot. The parliament supported the allotment of one-third of the total budget for military purposes. During the years 1863–1865 Romania purchased from French suppliers, in particular Godillot, 47,600 rifles, 700 pistols, 5,300 carbines, 4,000 muskets, 12,000 sabers, 200 Enfield carbines, and 48 cannon. Simultaneously, Cuza founded the national arsenal and armament factories.[21] In his own assessment of his achievement in building Romanian armed strength, Cuza reported to Napoleon III in 1865, "At the time of my election the principalities possessed no more than four or five thousand Russian rifles . . . and about ten cannon of no value . . . They possess today seventy thousand rifles . . . and my artillery numbers seventy-two."[22]

Cuza's fall from power and the arrival of Prince Charles of Hohenzollern in 1867 did not end the Romanian government's involvement in clandestine arms convoys into the Balkans and its efforts to acquire modern weaponry. However, the new political leadership did mark a switch away from France and the beginnings of German suppliers. In 1867–1868 the radical government of Stefan Golescu and later Nicolae Golescu devoted much energy to military affairs. Romania bought artillery from Krupp, Dreyse guns from Prussia, and Peabody rifles from the Providence Tool Company in Rhode Island. The French military mission continued its work until 1868, and Prince Charles also supported a Prussian military mission.[23]

In 1868 an intricate plan unfolded to bring in arms to the region, surreptitiously involving Russia, Prussia, and Romania. Arms of Prussian origin had arrived by the usual overland route from Russian Bessarabia into the principalities. The first convoy, with ninety cases marked as "railway material," enjoyed a most unusual escort composed of gendarmes dressed as peasants and two companies of plain-clothes police, directed by a subprefect. The convoy avoided the main roads and all towns. After crossing the Pruth, the group exchanged their Russian customhouse markings for those of Moldavia. An individual in Russian uniform, but said to be a Prussian, supervised the transport, and an employee from the Ministry of Finance in Bucharest received the cases. The local provincial governor inadvertently exposed the operation when he opened two of the cases and found they contained guns, cartridges, and bayonets. The governor then telegraphed St. Petersburg for instructions. With the secret out, it was claimed that the cases formed part of

50,000 rifles voted by the chambers for arming regular and frontier troops in Moldavia. "Upon being asked why arms manufactured in Prussia had been sent through Russia by a long and expensive route rather than by the Railroad through Galicia and the Bukovina, the Prefect replied that perhaps as his government had already got some guns through Austria they may have wished that these should pass by Russia to avoid the appearance of a preference for the former."[24]

It seems unlikely that Romania was the ultimate destination. The Romanian army numbered only about 15,000 men, whereas the arms shipments amounted to almost three times that amount. The Romanian government had already openly introduced 30,000 rifles from Austria, so why the circuitous route for this shipment?[25] In addition, Turkish authorities detained a Prussian vessel supposedly carrying cotton goods into Romania. In fact, the goods in question proved to be gunpowder. The French consul concluded that Prince Charles's government was clandestinely introducing arms to Bulgarian revolutionary movements. The British noted that "attempts to introduce needle guns as railway material and gunpowder as cotton goods do not inspire confidence in this Government."[26]

Following Russia's victory over the Turks in 1878, Bulgaria gained status as an autonomous principality within the Ottoman Empire, but in practice it functioned as a Russian satellite. Accordingly Russia took the initiative to organize the Bulgarian armed forces and to supply them with artillery, rifles, powder, and cartridges. At first the tsarist government provided vintage surplus arms, such as Krinka rifles left behind by the Russian army. By 1880, however, Bulgaria received new Berdan rifles as well as three batteries of new Krupp artillery pieces, courtesy of Russia. The artillerymen also received Smith and Wesson revolvers, and the cavalry obtained Winchester repeating rifles. Since the Bulgarian Army had been organized on the Russian pattern and provided with artillery and rifles from that empire, it came as no surprise that Bulgaria derived its stores and ammunition, cartridges, and powder from Russia. By 1885 the Bulgarian regular infantry were armed with Berdan Number 2 rifles, and the artillery consisted of Krupp guns—model 1876, captured from the Turks—or old pattern Russian guns.[27]

In addition to arming the Bulgarian principality, Russia looked after Eastern Rumelia as well. The Rumelian government was to be equipped with Berdan Number 2 rifles because the Russians considered it one of the best systems in the world. Russia offered two methods for Rumelian purchase. Either Rumelia could buy 35,000–40,000 rifles with cartridges at the regular state price from the Russian government, or Russia could make available one of the Russian arms plants for the Rumelians to make their own order according to commercial conditions, and the St. Petersburg cartridge factory could provide the cartridges. The first method would be more advantageous to the Rumelian government because it was faster and cheaper, and the Russian government could guarantee the work at its own factory. The Rumelian government took up the Russian offer and purchased the guns at 18 rubles apiece from the Tula factory. Ultimately Rumelia bought 40,000 Russian rifles for 225,000 rubles. Payment terms allowed for five years, and the first Rumelian payment for 5,000 rifles was made on September 15, 1883.[28]

In Ethiopia the impetus for the arms trade sprang from the rivalry between the regions of Shewa and Tigre and their rulers' competition for arms in order to dominate the Ethiopian state. The new emperor of Ethiopia, Tewodros II (1855–1868), attempted with difficulty to acquire arms and technical know-how from Europeans. Tewodros knew he needed to build his firearms capacity, but free access to imported weapons proved extremely difficult. Arms imports from the north had to pass with the approval of the Ottoman government, which controlled the city of Massawa. With the aid of Greek arms merchants, Tewodros had obtained only limited quantities: four cannon and one hundred rifles from Said Pasha of Egypt. The emperor was not the only Ethiopian leader seeking arms. In 1860 a rebel, Niguse, was negotiating with the French to acquire weapons of his own. He received 450 muskets and petitioned for French artillery. The emperor wanted English firearms, but these were prohibited by the Turkish blockade at Massawa in 1860.[29]

By the early 1860s Tewodros endeavored to acquire technical assistance from outside to manufacture his own armaments. He relied on the work by some Swiss Protestant missionaries to make cannon and mortars, despite their lack of expertise. Still looking to import firearms from the British, the emperor grew frustrated with the coolness of the Foreign Office to his requests. In 1864 he sent a German craftsman to Britain in hopes of arranging directly for English gunpowder makers and other weapons specialists to come to Ethiopia. He also sent dispatches to Queen Victoria, arguing for an English–Ethiopian Christian alliance against the Muslim Ottomans. After he was met with silence again, Tewodros decided to get British attention by imprisoning their diplomatic personnel along with other Europeans.[30] He planned to detain the captives until the military artisans and teachers were sent from England to teach the Ethiopians. As Tewodros explained in a letter to the British civil servant, Hormuzd Rassam, in April 1866, "What I require from you . . . is that you send me a cannon maker, a gunsmith, an iron smelter, a sapper, and a gunner. By the power of God, let all these craftsmen come to me together with their equipment, so that they may return after instructing me."[31] The Foreign Office sent Colonel Merewether out with the machinery and the craftsmen, and they arrived in Suez on December 5, 1866. However, Tewodros would not release his captives and the artisans were advised to return to England.[32]

Confronted with Tewodros's intransigence, the British sent the Napier expedition to chastise the emperor. The force landed near Massawa in January 1868 and moved into the hinterland. In April 1868 Tewodros was crushed in battle and later committed suicide. As a means to secure his supply lines, Lord Robert Napier had negotiated an understanding with the future emperor, Yohannes IV (kassa of Tigre) that in exchange for safeguarding the British rear, Yohannes would receive money and weapons. In all, Yohannes gained 6 mortars, 6 howitzers, 725 muskets, and 130 rifles. More important even than the weapons, Napier granted Yohannes 354,230 rounds of small-arms ammunition, 28 barrels of gunpowder, and 585,480 percussion caps. The value of the supplies amounted to approximately £500,000.[33] Thus, one Ethiopian ruler was punished for trying to acquire quantities of weapons from the British, while a future Ethiopian emperor was awarded large caches of arms from the British for his assistance.

The desire of Ethiopian leaders for the arms trade contributed greatly to conflict between Ethiopia and Egypt. In 1868, the Ottoman sultan transferred the city of Massawa to Egyptian control. As the new sovereign power, Khedive Ismail of Egypt clamped down harder on the arms imports into Ethiopia. The khedive initiated a blockade on arms and ammunition, although a contraband trade in weapons continued at an estimated 4,000 rifles a year until 1871. In 1872 the Massawa governor suppressed the export of arms into Ethiopia entirely. Along with territorial disputes, the arms blockade gave impetus to an armed conflict between Egypt and Ethiopia. Yohannes's resounding Ethiopian victories at

Gundet in November 1875 and Gura in March 1876 brought him not only glory, but also a windfall of modern weapons. Yohannes captured 20,000 Remington rifles and 25 to 30 cannon from the Egyptians. The terms of peace between Egypt and Ethiopia allowed that Emperor Yohannes could import 50 pounds of powder, 10 muskets, and 5,000 caps annually for his personal use.[34]

Just as Yohannes's predecessor had had to contend with a political rival and arms imports, the emperor, too, faced a challenger. In this case the rival was Menilek, King of Shewa. Menilek also clearly understood that no longer could a leader contemplate ruling Ethiopia without firearms. In June 1865 he, like Niguse before him, had looked to the French but had been stymied by the British. In 1872 he sent an Ethiopian priest to France and Italy to have them initiate arms traffic. In 1875 Menilek tried again, and this time the Frenchman Pierre Arnoux provided equipment to produce gunpowder. Around the same time Menilek also appealed to the khedive, who responded by sending 500 rifles and an artillery piece to Shewa in July. Khedive Ismail hoped that by arming Menilek he could put pressure on Yohannes, but the Egyptians soon placed Menilek under the same restrictions as Yohannes. So, like Tewodros, Menilek turned to Britain for arms. In 1878 he attempted, without success, to convince the British that he needed weapons to put down the slave trade. Britain had no interest in contributing to the potential threat to Egypt, and therefore the British rebuffed Menilek and Yohannes both.[35]

Menilek finally found a sympathetic response in Italy. Following Menilek's priestly envoy, the Italian Geographic Society had arranged an expedition that arrived in Shewa in 1876. The Italians came bearing gifts of 200 Remington rifles for King Menilek. In reply to the king's insistence that he required more, the Italians promised 2 mountain guns and 11,000 Remington cartridges. In 1879 Italy, through Count Pietro Antonelli, offered to open the route from the port of Aseb and deliver more arms. The specifics entailed 2,000 Remington rifles from private merchants. Menilek signed the agreement in 1881, and by April 1883 Antonelli had delivered the weapons.[36]

By the early 1880s Italian and French competition to supply Menilek with secondhand arms had created an accelerating weapons trade. Italian officials wanted Menilek's aid against Yohannes, whereas the French interest grew from private traders. Louis August Brémond, a French arms merchant, visited Shewa in January 1880 and lobbied Menilek to divert caravans from the British port of Zeila toward Obok. Menilek responded favorably to the French offer and advanced Brémond cash and ivory in exchange for 5,000 Remington rifles and some machinery to be landed at Obok. The French traders succeeded in launching their foray into the Shewan arms traffic through Obok, but since there existed no French-controlled port yet, these plans did not come to fruition until France formally occupied Obok in July 1884. A French trader, Paul Soleillet, shipped war materials into the port at Obok and established a trade depot there. By the end of 1882, Menilek had purchased 200 breechloaders and 2,000 muskets from Soleillet.

The Italians did not sit idly by. Antonelli brought in 5,000 Wetterly rifles through Aseb. In January 1883 some 5,000 rifles were off-loaded at Obok, and the next month Léon Chefneux, a business associate of Soleillet, and Eloi Pino brought an additional 1,300 rifles and 2 cannon. Not to be outdone, Antonelli and the Italians committed to deliver 50,000 Remingtons and 10 million cartridges over the next decade and 4,000 Wetterly rifles with 400,000 cartridges immediately.[37] In 1883 Antonelli commented about Menilek, "He asks for nothing but arms, which noticeably are the only currency in use in East Africa."[38]

By 1884, British officials had grown concerned about expanding French interests in the Red Sea. Fearing the French, the British reached out to the Italians. Already among the tribes east of Zeila, Ibrahim Abu Bakr was sounding out the French about selling a site. In view of the anticipated extension of Italian influence over the western shore of the Red Sea, Britain took the opportunity to cooperate with the Italian government and to restrict the import of arms through ports that might come under their control. In general, the unrestricted distribution of arms and ammunition south and east into territories occupied by the Oromo and Somalis posed a danger to travelers on the trade routes, as well as an encouragement to intertribal fights. Specifically, the ports in the Gulf of Aden would be jeopardized if an unlimited supply of firearms were made available to the Somalis.[39] For these reasons, Britain encouraged Italy to take control of Massawa from Egypt to block the French penetration to the Nile basin, effectively double-crossing Yohannes by letting the Italians in.

Italian Minister of Foreign Affairs Pasquale Mancini concurred with the British that it was necessary to restrict commerce of arms in the Red

Sea. But Italy called attention to the extension of French influence resulting from the large importation of arms through French Obok, which was carried on with great activity and had reached Shewa. Italian concerns arose that the proposed restriction might inadvertently give French merchants a monopoly in the arms trade. Mancini thought it was very important to avoid throwing the arms traffic entirely into the hands of French merchants. He opined, "The influence of that trade over the savage inhabitants of that coast was very great, and the country that supplied the natives with the greatest number of arms was the one which would have the greatest influence over them."[40]

Soon after the Italian annexation of Massawa, Yohannes perceived the Italians as the new threat. On February 5, 1885, Italian forces landed at Massawa with the blessing of Britain. When the Italian envoy came to make a treaty, Emperor Yohannes insisted he first obtain accreditation from Queen Victoria. As he protested to the British, "When I said this, the Italians were angry, and stopped the import of guns into my country . . . I wish you to give me Massawa."[41] Yohannes again complained bitterly to Queen Victoria in October 1887: "It was agreed that no arms should be imported into Abyssinia without my authority; but the Italians allow arms to be imported by any way, and owing to this my country becomes rebellious. Of old Massawa was mine. First Ismail Pasha and now the Italians wish to make war."[42]

Menilek spied a new opportunity to advance his own cause by undermining Yohannes with Italian aid. To that end, the Shewan king strove to maintain friendly relations with Italy, and consequently his friendship yielded thousands of Italian arms transported to him via Aseb. For their part, the Italians tried to undo the negative impression caused by their occupation of Massawa by stepping up arms deliveries to Menilek. In July 1886 they brought 4,000 Wetterly rifles and 400,000 cartridges. While Yohannes's general, Ras Alula (1847–1897), moved against the Italians, Menilek seized the opportunity to occupy Harar in January 1887 and to install his cousin Ras Makonnen (1852–1906) as the new provincial governor. The Italians endeavored to squeeze Yohannes and Ras Alula through the imposition of an arms blockade starting in May 1887. Simultaneously, Italy stepped up the flow of arms to Menilek, offering him 5,000 Remingtons within six months; by November 1,000 had arrived.[43]

With Makonnen's support, Harar became the heart of the Shewan arms trade, further strengthening Menilek's hand. In March 1887 Abu Bakr solicited Menilek from Obok to redirect the Harar trade to the area of French control on the coast. As Menilek's lieutenant, Makonnen took the initiative and had the Abu Bakr family facilitate a caravan route through Somali territory to the potential French sphere west of Zeila. In turn, the Gulf of Tadjoura and the port of Obok, under French control, now became wedded to the Ethiopian arms traffic. By September 1887, the arms trade figured prominently in the commercial activity in Harar. Some estimates credit Menilek with possessing 30,000 rifles by the end of 1887. By October, Makonnen, with the assistance of the Abu Bakr clan, had arranged for all caravans to use Djibouti just as soon as the French claimed it free of British influence.[44]

By the end of 1887 Ethiopia was well provided with firearms. Italian Major Piano put the armament of one hundred Ethiopians at an average of 45 percent armed with Remingtons (rifles and carbines), 5 percent other breechloaders, 20 percent muzzleloaders, and 30 percent spear and shield. He reported: "The favorite type of rifle is the Remington called by the Abyssinians the 'Senadir' (corruption of Snider), and great efforts have been made by the King and Ras Alula to get a large quantity of these rifles into the country and to arm all their troops with them . . . Needless to say, there are no manufactories of rifles in the country; the repairs are done by a few European workmen, but there is hardly a native with any mechanical skill at all."[45] A lack of a reliable supply of ammunition remained a serious drawback for Ethiopian forces. Ammunition formed the chief article of trade between the interior and the coast. According to Major Piano, the origin of the war with Italy traced to the Italians placing a heavy duty on arms and ammunition imports at Massawa. Another obstacle to obtaining a uniform pattern of arms and ammunition was that each trader on the coast supplied his customers with one particular pattern of breechloader other than the Remington; when the ammunition supplied with a rifle was exhausted, an Ethiopian was bound to go to that particular trader to get more. By this means, the arms merchants acquired a stranglehold over the supply of their particular ammunition. Powder and bullets for muzzleloaders were made in the interior. In Piano's assessment, besides small arms, Yohannes had forty pieces of artillery, including thirty Krupp guns taken from Egyptians in 1876.[46]

The last major development with profound implications for the arms

trade into Ethiopia occurred late in 1887 when the French moved to secure themselves in Djibouti. The French hoisted their flag at Djibouti on
December 1, 1887, and they proceeded to establish a free port.[47] British
consul E. V. Stace strongly suspected, correctly as it turned out, that the
sons of Abu Bakr Pasha had promised to arrange for the introduction of
firearms through the new French port. Stace worried that, "of course,
foreign adventurers will be found ready and anxious to aid in such
ventures. Such an arrangement would, of course, be viewed with the
highest favor in the interior, and its authors cordially assisted in their
objects."[48]

During 1888 both the French and Italians continued to sell arms and
ammunition to Ethiopia. The French took advantage of their new position in Djibouti to divert trade from the British in Zeila. Using an especially large consignment of 8,000 rifles as an incentive, Armand Savouré,
a French arms merchant, managed to persuade Menilek to avoid Zeila by
sending caravans to Djibouti. By March Savouré had begun to ship
weapons straight from Europe, bypassing the British in Aden, and proceeding without interruption into Djibouti. His first delivery from Djibouti to Harar arrived in October. Meanwhile, the Italians shipped 4,000
Remingtons, 200,000 cartridges, 5,000 other types of rifles, and 400,000
Wetterly cartridges to Menilek. For his part, Menilek hinted that he was
on the verge of breaking with Yohannes, and proposed that the Italians
provide him with a loan to purchase 10,000 more Remingtons and
400,000 cartridges. Reluctant to make the loan, but very eager to turn Menilek actively against Yohannes, Antonelli instead sent 5,000 rifles and 1
million Wetterly cartridges in January 1889.[49]

In 1888 the Italians wanted to enter an agreement with Britain and
France to restrict arms importation into their respective possessions on
the Red Sea and Somali coasts.[50] All three powers signed a convention
stipulating that arms should be imported only on license in June 1888,
but the Italians and French had decidedly different interpretations of the
convention's meaning. Italy viewed the convention as applying exclusively to private traders. Since they used government envoys rather than
private merchants to convey their arms to Ethiopia, the Italians believed
they were already in compliance with the agreement and could go about
their business unhindered.[51] French Minister of Foreign Affairs Emile
Flourens acknowledged the possibility of preventing the acquisition of
European precision arms by the neighboring "chiefs of small and bar-

barous countries," but since Yohannes and Menilek were "powerful independent princes, possessing considerable military forces," the Europeans could not prevent their importation of arms. Flourens referred to
the difficulty of preventing commerce in arms with any state or sovereign that desired arms and was able to pay for them. He observed that
Yohannes's troops were equipped with repeating rifles that were not of
French manufacture or obtained through French agency, and he concluded that if the princes did not obtain war material from the portion of
the coast that was under French or English protection, they would get it
from Italian or other sources.[52]

Britain wanted France to restrict not only the importation of arms into
the Somali coast, but also their transit through French and British territories. According to the French, the prohibition on imported firearms
did not apply to rifles sold to King Menilek. In the meantime, it was difficult to say how far Ethiopia had been inconvenienced by the British
blockade. France was doing its best to make it as painless as possible,
and 20,000 rifles came to Ethiopia by way of Obok.[53]

Disagreement also broke out between France and Britain over the
meaning of the February 1887 agreement relating to the import of arms
and gunpowder. French prime minister René Goblet said that as matters
stood, the two governments were apparently not in agreement. It might
be a matter of grave consideration whether either government would
facilitate the armament of the semibarbarous states adjacent to the protected territory of both. However, he doubted the possibility of preventing the rulers from obtaining arms through territories between them and
the coast. It further appeared to Goblet that it could not have been the
intention of either government to include the exportation of arms in any
prohibition placed on their importation. For instance, Goblet mused, if
the French government were to establish arms manufacturing within the
Somali territory under its protection, could it be reasonably required by
Britain to prohibit or interfere with the export trade of that product?
Lord Robert Bulwer Lytton replied that such a step on the part of France
would certainly be of a most extraordinary character. The French argued
further that the arms sent from Marseilles to Shewa had been promised
by the French government to King Menilek before the February Agreement. Therefore, the French claimed they could not prohibit this delivery. Lord Robert Salisbury instructed Lytton to tell the French that the
quantity of rifles seemed very large, and to inquire whether the promise

made to Menilek could be satisfied by some smaller quantity, less likely to have the appearance of military assistance to a potentate who was still engaged in hostilities with Italy.[54]

By February 1889, the French were threatening to withdraw from the agreement with Britain. According to the French view, the only ones benefiting from the agreement were the Italians, who the French claimed were carrying on a brisk trade with Harrar and the interior. At the same time, British consul Stace was perplexed about how to answer merchants who applied for permission to take in arms for caravan protection to Harrar-Shewa. If he refused permission, those in Harrar-Shewa would import their arms through the French in Djibouti, and thus divert the Zeila trade to the detriment of British interests. Eventually, the Italian government accepted, in principle, the restriction of imported arms of Italian, British, and French protectorates on the northeast coast of Africa, if the restrictions applied only to private commerce.[55]

Among the three powers that had made the agreement, only Britain seriously enforced the measures. Consequently, the British in Zeila found themselves sitting on the sidelines while the Italians and French conducted a brisk trade in arms and ammunition. The British consul in Zeila, Colonel Stace, could see the futility of the British position as Zeila declined and Djibouti rose, and thought that Britain should stop excluding itself from the arms trade. Under the existing arrangements, in which arms traffic was allowed from Tajourra and Djibouti and prohibited from Zeila, the greatest encouragement was given to the slave trade, chiefly in the hands of the sultan of Aussa (Dankili) and the family of Abu Bakr. Stace argued, "If arms are freely imported via Zeila, the occupation of the Abu Bakr family will, in great measure, cease, and a serious blow will be dealt to the slave trade. Moreover, it will be a decided check on Djibouti."[56] The Shewan occupation of Harrar gave Menilek alternative routes, but he was unable to import arms via Zeila; he was in the hands of the Abu Bakr family, who collected the camels and introduced rifles via Djibouti.[57]

Although Menilek primarily devoted his efforts to importing more weapons and ammunition, he also pursued the means to manufacture cartridges. In 1888 the king had sent Alfred Ilg, a Swiss engineer and faithful supporter, to Europe for equipment to make cartridges. Ilg reported to Menilek from Zurich in February 1888 that the British authorities in Aden told him they would not let any machinery through to

Ethiopia, since trade in rifles and cartridges was forbidden. Despite British determination not to let the machinery through under any condition, Ilg promised to send whatever he possibly could.[58] In April 1889 Ilg was involved in setting up machinery he had imported for cartridge manufacturing in Dabra Berhan, and by June the installation was almost finished. Ilg anticipated cartridge production would begin shortly.[59] He reported to Menilek, "I am beginning the production of cartridges by the will of Christ, and I will send samples to His Majesty . . . The rifles will be good, and they will be serviceable to Your Majesty."[60] Soon thereafter Ilg sent the good news about the cartridges and some samples in a box as proof of the successful production process. Ilg pointed out to Menilek the extraordinary efforts he had undertaken in the king's service, noting:

> For my part, I exerted myself a great deal on behalf of Your Majesty. Unafraid of imprisonment and confiscation when coming from abroad, I brought the machine through for you. I count for a trickster before the British. Henceforth I shall have no access through their ports, for I am an enemy of many people . . . I will in accordance with Your Majesty's order repair the cartridge gadget of Monsieur Antonelli . . . It would be good if you could assign six workmen and 20 assistants to the cartridge production, for much cannot be produced if the number of workers is not increased.[61]

Ilg also manufactured the cartouche for Antonelli's Snider gun and improved it.[62] Nevertheless, ultimately domestic cartridge production did not dramatically increase, even with Ilg's efforts.

Menilek had not yet shown inclination toward the British or French. Ilg urged Menilek not to trust Italy and to turn more toward France. Even though Vincenzo Ragazzi, an Italian doctor, was bringing 1,000 Remingtons and 150,000 cartridges, Ilg expressed wariness about Italian designs. He advised Makonnen in June 1888 that "Italy is not a friend to the Amhara. And now, Italy has quarreled further with the ruler of Zanzibar. They are looking everywhere for territory. The English and the French are not evil today. The French talk about letting rifles through. They claim not to hamper the importation henceforth."[63] For their part, the British worried that Menilek would fall in with French designs, if he could obtain firearms through that nation. Meanwhile, fifty camel-loads of rifles left Djibouti for Harrar, and at least two merchants were bringing 2,000 rifles each from France.[64]

As Ethiopia's relationship with Italy changed, so did the arms trade. In 1889 Yohannes died fighting against the Mahdists, and Menilek assumed the position of emperor. On May 2, 1889, Menilek signed the Treaty of Wichale with Italy. The Italian version of the treaty bound Ethiopia to Italy as a protectorate in Article 17, although the Amharic version contained no such provision. France and Russia protested the treaty. Neither power recognized the Italian protectorate since, under the terms of the Berlin Conference of 1885, Italy lacked effective occupation of the territory or a formal proclamation of protection. By November 1889 Italy had entered friendly relations with Ethiopia's new emperor, and the Italian position changed. It was no longer convenient to prohibit arms imports to Harrar. The practical object was to prevent muskets and powder reaching hostile dervishes. The Italian change meant that the tripartite declaration on arms imports should be abandoned, according to the British India Office. With Ethiopia coming under Italian protection, the interdiction on arms through Zeila was unnecessary, and it served only to give France a monopoly in the trade to Shewa through Djibouti.[65]

Menilek actively manipulated the Italian protectorate to acquire arms. In 1889 he sent Makonnen to Italy to acquire weapons. Makonnen entered into direct negotiations with Italian officials with dramatic success. He managed to buy from the Italians 10,000 Remington rifles, 500,000 rounds of ammunition, machine guns, and artillery. These materials came from Italian arsenals, and the Remington rifles arrived in Ethiopia by September 1890. Not only did Italy open the tap and let the arms flow, but the Italian government also intervened to knock down the price. Typically the rifles cost $30 each through private traders, but thanks to subvention by the government Makonnen paid only $5 apiece.[66]

Since the restriction of imported arms through French and English protectorates was no longer in complete force, the question of British participation in the arms trade arose. Stace wanted to know whether arms and ammunition genuinely for the government in Harrar could now pass through British-controlled Zeila without restriction. In Salisbury's view, under the General Act of the Brussels Conference:

> As regards the special case of Harrar, I have to observe that Abyssinia has adhered to the General Act, and is, therefore, under it, entitled to receive arms for the use of its authorities, though not for sale, apart from its right to do so under the Protectorate Treaty with Italy. Unrestricted trade in arms through Zeila is clearly barred by the provisions

of the General Act, but whenever the Act is ratified, the transit of arms to Harrar for the use of the rulers of the country, must, on application of the Italian Government, be sanctioned.[67]

Along with the Italians, the French also encouraged the arms trade, and the British eased their restrictions in Zeila. By 1891 Djibouti and Obok were becoming the main centers for weapons distribution. Even machine guns found their way from Djibouti to Harrar.[68] As far as the French in Djibouti were concerned, Ilg said, "As to firearms, do not worry as long as we are there. We have not hampered the import until this day; we do not intend to hamper the import from now onward either."[69] In November, an agent of Makonnen requested permission to send 100 rifles and 20,000 cartridges through Zeila. Given the importance of friendly relations with Makonnen for the British protectorate in Somaliland, and since large quantities of arms were passing to Harrar through Djibouti in any event, the British complied with the request.[70] Makonnen replied, "We became very glad and thankful to the great and illustrious Government and to your Excellency for the permission granted for the importation of guns and ammunition via Zeila. The permission granted shows the friendship and amity which the illustrious Government bears for our Government . . . We will now at once write to His Highness John Howi Menilek about the pleasing news, so that he might be glad."[71] To demonstrate his pleasure, Makonnen sent three lions to Aden for presentation to Queen Victoria. He further asked that 200 rifles and 20,000 rounds be allowed to pass through Zeila. By March, his request had increased to 4,000 rifles.[72]

British officials became more and more alarmed at the spillover effect of the Ethiopian arms trade on their coastal possessions. As the arms trade between Djibouti and Harrar, which had existed since 1889, increased in volume, British concern grew that Ethiopia's acquisition of large quantities of arms "may constitute a serious menace to the security of our Somali Coast Protectorate."[73] Since France had not acceded to the Brussels act, the British saw no need to inquire about the French arms trade through Djibouti. The India Office began to express the need for Britain to engage in some portion of the arms trade as a matter of self-defense, since the Ethiopians raided the Somali tribes of the British protectorate with impunity, but ports of the British protectorate continued the arms trade restrictions.[74]

By 1893 Italian and British reports observed the activities of France to substitute its influence for that of Italy in the region west of the Red Sea, and its importation of considerable quantities of arms and ammunition. At first the Italians ascribed the more aggressive posture of the French to the excess of zeal and personal ambition of their officials and agents in Djibouti, rather than to any deliberately hostile policy on the part of the French. By 1895, however, Italian suspicions had grown, and they had come to believe that France was trying to undermine Italy's influence with Menilek. Baron Blanc saw a connection with the Russian expedition into Abyssinia. He feared that although the alliance between France and Russia did not count for much in Europe, it had found, on the borders of the Gulf of Aden a field for common action against Britain and Italy. According to Italy, France's action to allow arms to be transported to Menilek was destroying the Italian agreement with Britain of May 5, 1894, to establish a Franco-Russian protectorate, and the Russians were helping by bringing in arms under cover of their expeditions across Somaliland.[75]

The Italians were not alone in complaining about French participation in the arms trade. Private French traders also bemoaned the stiff competition offered from Djibouti under Govenor Léonce Lagarde's administration in Obok. In 1893 the Djibouti colonial government allowed Makonnen open access to purchase 1,000 carbines, 500 Gras rifles, and 150,000 cartridges from government stores. Savouré observed with despair that in 1893–1894 Lagarde had sold 2,000 Gras rifles at an incredibly low rate of $6 each, thereby underselling the private merchants by more than two-thirds.[76]

As it became evident that Menilek intended to denounce his treaty with Italy,[77] the Italians looked to Britain to establish a common front to counter Menilek's growing assertiveness. In particular, Italian officials enlisted British support in continuing the prohibition on the arms trade from British ports. In a secret report to the Italian Ministry for Foreign Affairs in January 1895, the Italians alleged that because England seemed indifferent toward the Italian protectorate in Ethiopia, Menilek could scheme as much as he pleased with Italy's enemies. By the Anglo-French agreement of February 29, 1888, neither would annex Harrrar and both would prevent arms imports into their territories. Even though the French were supplying arms, Italy requested that Britain withhold permission for Makonnen to import arms via Zeila. While acknowledging that such a prohibition would be illusory, as the governor of Harrar could procure arms from the French ports, the Italians believed that "all the same it should be remembered that the importation of arms into Abyssinia from an English port has a very different significance in the eyes of those populations from which it would have if the arms were imported from the ports of our adversaries, as it appears to show that England is not the friend of Italy."[78] The Italians wanted to intercept Makonnen's arms shipment, and they requested that Britain prevent the arms importation from Aden under the Brussels act, and Britain agreed.[79]

The French continued to complicate matters for the Italians. They had stated to Italy that France had prohibited supplies and arms to native tribes on the coast, but had made no such rule as regards Emperor Menilek and his officers, as he had acceded to the Brussels act. The French also denied that Ilg had any connection to the French government. The Italians found it somewhat strange that the French government—having refused to recognize the adhesion of Menilek to the Act of Brussels when it had been ratified by Italy as the protecting power of Ethiopia—should now plead that adhesion as an excuse for permitting the supply of arms to Ethiopia when they were endeavoring to throw off the protectorate.[80] Meanwhile, Ilg urged that Ethiopia look actively to the French. He met with Lagarde in Paris in 1895. According to Ilg, Lagarde said, "Only fine words and letters come from Ras Makonnen, but nothing fruitful. Ras Makonnen allows exports of articles through Zeila, but he requests me to let weapons through on my side." Ilg further warned Makonnen, "Will you not consider what can happen? Can weapons and cartridges be imported through Zeila when the French port is closed? There is no one who has been so useful as the French. There is no one who has profited less than the French. What will become of us if we do not agree with the French? . . . The greatest enemies are the Italians and the English. Why do you help these? . . . Please, be in harmony with Monsieur Chefneux. He labored greatly for you."[81]

With all the uncertainty about European intentions, Menilek sought to test British support for his arms supply. Ilg had warned Menilek that the Italians intended "to isolate the Ethiopian government from all the other governments and direct it according to their wish. They claim that if they run into conflict with Menilek, they will reach an agreement with all other governments and hamper import of firearms and cartridges."[82]

The emperor dispatched Christo Moussaya as an arms agent to persuade the British to ease up on restrictions by threatening to go to the

French. Moussaya was one of three Greek brothers who had lived in Harrar and who were confidential and commercial agents for Makonnen. Moussaya arrived in Harrar to obtain a license to import into Zeila 660 arms that had recently landed via a French steamer. Moussaya indicated to British authorities that Menilek had come to believe that the French were his only friends and that his interests lay in holding to them alone. Moussaya related that everything was done to make Menilek believe that Britain supported Italy's plan to annex Ethiopia. According to Moussaya, Makonnen took another view and tried to impress on Menilek that France was not to be trusted. The French, jealous of Italian development on the Red Sea littoral, wanted to oust the Italians. Moussaya told the British that Makonnen had urged Menilek to place more faith in England. Menilek told Makonnen to produce proof by getting some arms through Zeila. Moussaya added that if the arms were sent as promised, Makonnen would be able to reestablish himself in Menilek's confidence, and the French influence would be checked.

Moussaya pointed out that hundreds of thousands of arms had already gone into Ethiopia, and many more that had been paid for were already in Djibouti and Obok, brought there by speculators. It was ridiculous to think that the few arms promised to Makonnen through Zeila would make any difference, he said. Moreover, if the arms were sent back to their consigners, there was nothing to prevent them being reshipped to Obok or Djibouti. Moussaya added that if the present consignment of 660 rifles in British hands were forwarded, it was possible that the balance of the 4,000 would never be shipped to Aden since Makonnen's views would be vindicated.[83]

Meanwhile, Makonnen expressed his own dissatisfaction with the British directly in a letter to the British consul on the Somali coast. "We ordered Kawaja Abu Sitta (Moussaya) to purchase for us some guns . . . Before the purchase we informed you, and you permitted the same. If we are not to get the guns their cost is due, and we will leave them. But, knowing that your Government and our Government is one, our friendship still continues, our property and yours is the same. We did not think the property of John Howi (King of Kings) would be detained without any cause, disturbing the friendship and peace existing between us."[84]

On the British side, frustration with Italian inconsistencies grew. In deference to the Italian protectorate over the Ogaden, on November 17, 1894, the British government offered to suspend the permission already granted to Makonnen to import some arms. The Italian government accepted the offer, but on December 5, the Italian vice consul actually imported a considerably larger quantity of arms, with ammunition, as a present from General Baratieri to Makonnen.[85] Finally, in 1897 the Italians dropped their opposition to the delivery of rifles detained in Aden, and the 660 rifles were delivered to Makonnen in April.[86]

As for Ilg's mission to Paris to forward without delay arms and ammunition via Djibouti,[87] Ilg reported to Makonnen in April 1895, "I found old-model Remington firearms, but I have not yet bought them. To save time I will borrow money and purchase and ship a big quantity."[88] This mission led to a major shipment of arms from Liège to Ethiopia conveyed by a Dutch steamer. The shipment—1,608 cases originally from Russia—included 15,000 Gras rifles and 25,000 others of French and Russian origin. Additionally, 5 million rounds of ammunition and artillery shells were on board. The principals who had arranged the cargo, Chefneux and Ilg, had borrowed the money in Europe to procure the materials and charter the ship. The Italians intercepted the steamer in the Red Sea on August 8, 1896. The weapons did not reach Ethiopian hands until 1897.[89]

The steamer shipment represented only the most prominent piece of the Ethiopian arms trade. British intelligence in Brussels discovered that 7,000 rifles had found their way from Liège to Antwerp to Marseilles to Obok. Arms were purchased by an Armenian, Serkis Terzian, who was also believed to have procured arms in Belgium for insurgents in the Ottoman Empire. Since 1892, between 80 and 100 camel-loads of arms and ammunition had departed to Harrar and Ethiopia almost every week. In return, the traders received payment in gold and ivory. Native government officials said that the French government in Djibouti also sent presentations of arms and ammunition to Menilek and Makonnen. Despite the protests of the Obok governor and the assurances of France, the arms transit was not stopped. The Italian vice consul at Aden affirmed that Lagarde, the governor of Obok, actively participated in the trade. An employee of the French steam line Messageries Maritimes told the *Djibouti News* that a Russian vessel arrived at Obok at five A.M. January 28, 1896, and discharged a cargo of small cannon and 5,500 rifles.[90]

Italian intelligence confirmed a large purchase of arms and ammunition in Belgium on a vast scale for transmission to Africa (Ethiopia and Sudan), Armenia, and China. The principal centers of the trade were

Liège and Antwerp. Two groups of purchasers in France and Belgium supplied the arms and ammunition to Ethiopia. In France, the dealers consisted of Léon Chefneux and Ghinaud in Paris. The Belgians, operating out of Wettern, supplied primarily Gras rifles to Shewa. The Fonderie Royale de Wetteren, Societe Coopal et Cie, by Ghinaud's order, manufactured black powder for Menilek's army. In Belgium, Terzian served as the chief buyer of arms for Menilek's account. Based on estimates from suppliers in Belgium, Menilek had 300,000 military rifles, mostly old Gras, and some Spanish Remingtons.[91]

Menilek had achieved his goals to gain control of Ethiopia and to hold back the forces of European imperialism through his astute manipulation of the secondhand arms trade. With Italian arms he managed to challenge Yohannes and then assume power as emperor. Under an Italian protectorate Menilek amassed still more weapons from the Italians and from the French. Having ably employed that firepower against the Italian invasion forces at Adwa in 1896, Menilek threw off would-be Italian control and enthusiastically embraced a greater arms trade from French Djibouti. A key element in Menilek's success derived from his playing on Italian imperialistic assumptions that they were gaining influence over Ethiopia by providing the emperor with arms.

The eastern question and the scramble for Africa shaped the secondhand arms trade. In the eastern question the Romanian lands under Cuza served as the nexus of arms smuggling, and common cause was made with Serbia to assert the right of national self-defense through the arms trade. Here the Russians and French actively engaged in clandestine and open rifle supply as they carried out their own rifle modernization at home. On the international level, the scramble for Africa prevented the formation of a united front of France, Britain, and Italy to inhibit the arms trade to Ethiopia. Various understandings and negotiations, from bilateral to tripartite, to the Brussels act, failed to stem the flow. Italy, the primary supplier in 1881–1896, tried unsuccessfully to use the arms trade as a tool to create an Ethiopian protectorate. The Italians succeeded only in providing Menilek with the most weapons with which to resist Italy's territorial ambitions. The French officially entered the picture in 1888–1896 largely through activities of their colonial officials in Djibouti, although French traders expressed much eagerness to play a role. At times France, too, prohibited the importation of arms for

Ethiopia when the French government needed to establish territorial agreements with Britain in 1886–1887 or when the outcome of the Italo-Ethiopian war was not yet certain.

At the heart of both sets of international problems lay a struggle by local rulers to receive recognition as legitimate states in the eyes of European powers that were more interested in controlling the local rulers. In the Balkans and Ethiopia the common course of action involved acquiring arms. Consistently, the British stood on the sidelines and lamented the flow of arms as unnecessary and dangerously destabilizing. Meanwhile, the French, Italians, and Russians unloaded their surplus arsenals into the hands of commercial traders or surreptitiously delivered the goods through state agents in hopes of acquiring influence over the local rulers and make them pawns in their larger game. The supplier states were each burned at one time or another by their policies, if not always as spectacularly as the Italians in Ethiopia. Just a few short years after the Russo-Turkish war of 1877–1878, Serbia, Romania, and Bulgaria all ended up in the orbit of Austria and Germany, thus leaving Russia and France with nothing to show for the influence they thought they had purchased through the arms trade.

3

Arms Trade Colonialism:
Ethiopia and Djibouti

Secondhand weapons continued to predominate within the Ethiopian arms trade. Nevertheless, the Ethiopian victory over the Italians changed the arms trade in East Africa quantitatively and qualitatively. The defeat of the Italians at Adwa placed a large quantity of small arms and ammunition and some light field guns into the hands of Menilek, such that the Ethiopian ruler was more than sufficiently armed to secure his independence. The French now achieved dominance over Italy in the Ethiopian arms trade, and the volume of arms increased dramatically. In previous years the French had poured arms into Djibouti, where arms sales had been permitted and encouraged. Consequently, Menilek succeeded in accumulating a vast supply of military stores and glutting the country with small arms. After 1896, the arms trade with Ethiopia became essentially a French business with secondary involvement from that nation's ally, Russia. Now Russian arms too, flowed into the country. In 1898 Tsar Nicholas II presented to Menilek 30,000 rifles and 5 million cartridges, and a Russian agent placed orders for eight Maxims and 250,000 Lee-Metford cartridges in 1900. The greater proportion of rifles in Ethiopia consisted of Remingtons, followed by Gras, Wetterly-Vialti, and Berdans. One firm alone at St. Étienne had supplied 350,000 carbines for Ethiopia, of which 150,000 arrived in March 1900. These were Gras Mousqueton carbines that had been discarded by the French artillery.[1] Increasingly, the weapons were not used by the emperor's armed forces to defend his realm. Instead, within the country Ethiopians were distributing the weapons to build their own patronage networks without Menilek's approval, or retrading the rifles to Somalis and others outside

Ethiopia for personal profit. In this way, the secondhand arms trade from Europe to Ethiopia gained a tertiary dimension.

British officials on the East African coast had been preoccupied for years with the arms trade at Djibouti and the danger it posed to British possessions. The question was brought more prominently to notice in 1901. During operations against the "Mad Mullah," the British discovered that a large portion of the rifles and ammunition captured from the enemy were made in France, and had been introduced into the country principally via Italian ports through native craft trade with Djibouti. The only effective remedy lay in closing the market. Unless the French government cooperated with the British, no matter what measures the British took, the trade would continue. Britain tried vigorously to persuade France without success. It had become clear that the necessary condition for any effective control of the arms traffic entailed the cooperation of Italy, France, and England.[2]

Any cooperation remained unlikely since all the participants in the business, Europeans and Africans, found some profit in the arms trade. All involved insisted that controlling the trade along its myriad routes exceeded their jurisdiction or capability to enforce. The French government always maintained that it was powerless to prevent arms shipments at Djibouti on vessels destined for ports outside the prohibited zone, as defined in Article 8 of the Brussels act. Local French officials had no interest in taking action that would incur the displeasure of the arms merchants; in some instances arms merchants and French colonial officials were one and the same. Considering that the commercial prosperity of Djibouti depended on arms traffic, any measures prohibiting, or even restricting, the trade would arouse the hostility of the Colonist Party. French officials insisted that the arms were consigned to Emperor Menilek, for whom arms and ammunition imports were authorized under the provisions of Article 10 of the Brussels act.[3] In 1906 the Italian minister in Addis Ababa had tried to get Menilek to prevent the illicit arms traffic in Somaliland. The emperor declared that he was unable to do this, and countered that if people who smuggled in arms were detected and stopped at other points on the Dankil and Somali coasts, the whole question would be settled.

Long diplomatic negotiations between the Italian and British governments to induce the French to restrain the illicit traffic had yielded little. The Italians wanted the French to guarantee that arms introduced into

Djibouti were truly consigned to the emperor, and not to Ethiopian chiefs and subjects. The decree of October 18, 1894, in French Somaliland allowed Ethiopian chiefs and subjects residing on the coast to acquire arms and ammunition by showing a written order from the emperor. This procedure was often abused or loosely enforced. The Belgian government also proved reluctant to discuss the arms traffic restrictions, due to the negative effects on the Liège arms industry.[4] In French Somaliland itself, three firms conducted most of the arms imports: Comptoir de Jibuti, Kevorkoff, and Garigue. An agent of the Parisian firm Société Francaise des Munitions in Djibouti reported that his company bought 1 million Gras rifles from the French government in 1906.[5]

Finally, Italy, France, and Britain signed an agreement on December 13, 1906, covering the Red Sea littoral, Gulf of Aden, Indian Ocean, and Ethiopia. Under the terms, arms destined for the Ethiopian government required documentation indicating the name, the number of arms, and the destination. The agreement coincided with the signing of the tripartite agreement by which Italy, France, and Britain divided Ethiopia into three spheres of influence.[6]

London and the colonial officials differed in their analysis of the problem and therefore the best solution. The British Foreign Office held Menilek responsible and proposed that he be urged to impose heavy fines on Ethiopian soldiers who sold their rifles to neighboring tribes. Britain also wanted the Ethiopian ruler to ensure proper disposal of obsolete rifles by the Ethiopian government, should it decide to rearm its troops. However, the British assessed the problem differently. For them, the question of regulation depended less on Menilek and more on the French authorities in Djibouti. They noted that Menilek had already issued proclamations forbidding the traffic. For example, Menilek had proclaimed that every man who had a rifle, with the exception of those received from the government, had to go with his chief and have it registered. Although British representatives did not want to appear to be treading on Menilek's status as an independent sovereign, they were not averse to threatening the emperor with a temporary embargo on all imported arms and ammunition. At the time, the Ethiopian government did not appear to have any intention to rearm its troops. Most soldiers had either Gras or old Russian rifles of the same bore. The only body of regularly armed men was the emperor's bodyguard of 1,500 men with Lee-Metford rifles.[7]

The formal agreement brought no change in Djibouti. British reports in 1907 noted, "The influences surrounding the French Minister are most unfortunate . . . both his Consul and interpreter are engaged in the traffic, and . . . it is openly pursued by the French Vice-Consul in Harrar."[8] In an attempt to expose official French participation in the illicit trade, British consul Sir Thomas Hohler gave Antony Klobukowski, the minister plenipotentiary sent from French Somaliland, the names of those engaged in the arms traffic, including the French vice consul at Harrar, M. Guigniony. According to Hohler, "Klobukowski expressed great surprise and regret at learning that M. Guigniony traded in arms, and promised to send strict instructions to him to refrain from doing so in the future." When confronted, Guigniony denied it, but the British vice consul in Harrar, John Gerolimato, produced a receipt for a Gras rifle purchased at one of Guigniony's houses.[9]

Concerns about the uncontrolled dissemination of arms into the hands of Somalis and Oromos had prompted Menilek to ask the three powers to ensure that rifles meant only for him passed through their territories. Greek and Armenian traders introduced weapons into Ethiopia declaring them to be for the emperor, but only some were destined for him. The rest went to Somalis and Oromos. The Somalis were increasingly arming themselves. In a recent encounter near Arana between two native tribes about seventy men were shot dead. In October 1907 the commissioner of Somaliland stated that, though the Somalis were not allowed to purchase arms openly in Djibouti, no restrictions were placed on Arabs or Danakils, who acted as agents for the Somalis. The British vice consul at Harrar, writing of the Hawiya tribe in the Ogaden, who were in revolt against the Ethiopians, reported that they had always been powerful, but had become much stronger after being furnished with a good supply of arms from Djibouti. He anticipated that all the Somali tribes would be so well armed in the near future that the Ethiopians would have great difficulty in preserving their rule in Harrar.[10]

Gerolimato, writing from Harrar, complained about the ongoing lack of cooperation from the French. After hearing that the French government in Djibouti had prohibited arms imports for Ethiopia, the British vice consul remained unimpressed. He wrote, "The only thing is to prohibit completely this import, if really there is the entente cordiale also for Abyssinia, for which I have some doubts, because in Abyssinia the Frenchmen want the whole for themselves and nothing for the others.

They put forward always that if the import of arms into Djibouti is prohibited the budget of this colony will be nothing, and then the metropole will pay all the expenses of Djibouti, but this is not a reason to arm all Ogaden, which is all the hinterland of the British Somaliland." In venting his frustration, Gerolimato proposed that the only way to make the French stop the import of arms into Djibouti was to put them in an analogous position. He wrote: "Should it be in my power I would oblige the French Government to stop really the import of arms into Djibouti. I would give permission, not to a merchant, but to some officer of reserve, to sell in Malta or in Pontellaria of Sicily, arms and ammunition just opposite of Algeria and Tunisia, and then at once, after one or two months, the French Government would accept and prefer to pay all expenses of Djibouti than have a market of arms in a point like that of Sicily or Malta, but this is not in our power."[11]

In December 1907 G. Colli, the Italian charge d'affaires in Addis Ababa, submitted a report to the Italian foreign minister detailing the state of the Ethiopian arms trade. Colli identified Guignony, the French consular agent in Harrar, as a participant in a consignment of 30,000 rifles shipped from Marseilles to Djibouti for the emperor. Colli evinced no undue alarm about the Ethiopian armaments. Given the present political situation, he deemed the idea of Ethiopian aggression against any of the neighboring powers as highly unlikely. The arms imports were essentially defensive, and arose from the conviction that the European powers, especially Britain and Italy, were looking for some excuse to intervene in Ethiopia. Only France escaped Ethiopian suspicions by providing Ethiopia with lavish quantities of arms.

Italian attempts to suppress the traffic in arms through Ethiopia and neighboring provinces only contributed to the suspicion. Colli wrote, "And it is especially for this reason that the loyal cooperation of the French Minister in our action against the traffic in arms would have been indispensable. His conduct, on the contrary, even if we admit that it has not been inspired by directly hostile intentions, has certainly been prejudicial to us, and has caused us serious embarrassments." Domestic uncertainty in Ethiopia also induced armaments by the emperor and his chiefs as each rendered himself more powerful than his neighbors. Colli continued, "Such armaments must certainly always be for us both dangerous and harmful, while for Ethiopia they represent a force and a guarantee for her independence so long as she remains united and faithful to

the Government of the Negus; but they would, on the contrary, be a terrible instrument of civil war and destruction should the Government
lack the necessary authority to repress disorders which will probably
convulse Ethiopia on the death of Menilek." Menilek knew that his
country had become saturated with arms. On occasion he had observed
that Ethiopia had too many arms and too few workmen. In practice,
whether motivated by personal suspicion of the European powers or by
a desire not to run counter to the wishes of all his chiefs, and especially
of those most loyal to him, Menilek continued to acquire arms and to
distribute them to his chiefs, thus increasing their power.[12]

While the Negus continued to acquire large ammunition supplies
from Europe, mainly through French merchants, Menilek seemed eager
to establish the domestic manufacture of cartridges. In 1907 an English
proposal to establish a cartridge factory in Ethiopia needed only Menilek's signature. Also proposed was a rifle factory that could produce
ten rifles per day.[13] The English syndicate for the cartridge factory was
led by William Knox D'Arcy. The sticking point in the concession was
Article 9, which stated that the Ethiopian government would purchase
up to 3 million cartridges yearly if the company was unable to make that
amount domestically. Hohler opined:

> I would like to add, confidentially, that if the present syndicate does not
> obtain this concession it is highly probable that some other, and possi-
> bly a foreign company will succeed in doing so. I venture to suggest
> that it is more desirable that, if the concession is to be given, it should
> be in the hands of a respectable British company of good standing than
> in those of any one else. His Majesty's Government are more interested
> than any other Power in repressing the arms traffic, and they would al-
> ways be able to control the proceedings of a British company, through
> the medium of this Legation.[14]

The draft called for Menilek to revoke the concession he had granted
to Sarkis Terzian on May 3, 1906, and give it to D'Arcy for fifty years. The
factory would begin production within two and a half years and make at
least 30,000 cartridges daily. The government would prevent competition and refrain from importing any caps, cartridges, or explosives from
abroad. The concession was to be designated the Ethiopian Government
Cartridge Factory.[15] With regard to the British cartridge factory applied
for by Julian Humphreys, although the emperor approved it, the conces-

sion had not yet been granted as of March 1908. The French representative, Klobukowski, reportedly tried to persuade Menilek that the deal was disadvantageous, although clearly the French interest was to protect its trade in ammunition imports.[16]

In 1909 Humphreys received the cartridge factory concession for 190,000 Theresian dollars (£18,000). Although the contract had been signed, no money had been paid, and the cartridge concession languished until 1910. Then the Ethiopian government granted a different concession to Humphreys for a cartridge factory combined with electric lighting and the purchase of machine guns. Under the terms, the Ethiopian government would pay Humphreys £19,000, of which £14,000 was to be expended on machinery and the electric plant. The remaining £5,000 was to order machinery. Humprheys was to become the general overseer for ten years and he would receive £1,000 for every 1 million cartridges produced.[17] As of 1913 not much progress had been made. According to Wilfred Thesiger, head of the British legation, the Ethiopians

> have spent and are spending large sums on the factory . . . Once the factory is in working order, the advantage to us would be that the output would then be limited, although amply sufficient for the real needs of the country, and the Government would have a clear interest in preventing the merchants from importing fresh stock until they themselves had disposed of the millions of practically worthless cartridges which they undoubtedly have in store, and would be quite willing to unload on the market by degrees as they filled up their arsenals from the factory. We should thus, for a few years at least, make the Government the sole dealer, and have the satisfaction of knowing that the cartridges they were placing on the market would be, owing to deterioration, vastly inferior to those which are imported from outside.[18]

The significance of this enterprise was not in the realization of this plant. In fact, the concessions never came to fruition. Rather, the support by the British for this arms business marked a departure from years of previous policy that opposed arms manufacturing within Ethiopia. Reflecting the growing sense of frustration in impeding the flow of weapons, the British position now conceded that the only way to stop the arms trade into Ethiopia was to manufacture the weapons inside the country.

British policy on the arms trade was caught between opposing interests. In 1908 the London Chamber of Commerce sent a letter to Edward Grey advocating for a British share in the arms trade. The merchants complained that Belgium sold more guns than Britain, and said that British merchants and manufacturers should at least be free from restrictions that did not affect Belgium.[19] At the same time, British officials in British Somaliland complained bitterly about the troubles caused by imported arms. According to their estimates, operations in Somaliland in 1902–1904 cost £8 million, and the annual military expenditure in Somaliland increased from £9,000 to £55,000 in nine years. The British officials added that the present force would not be able to deal with a serious threat. "This is a big price to pay for the illicit traffic in arms and ammunition,"[20] Montague Hornby wrote to Grey.

French official participation in the gunrunning continued unabated and unabashedly. H. F. Ward reported back from Harrar that he had encountered a caravan carrying 15,000 Gras rifles and vast stores of ammunition. According to Ward, Guigniony, now French consul at Dire Dawa, headed the caravan, and Ward bought twenty rifles from him personally. "Heavens!" exclaimed Ward. "If only people at home would realize the enormous bombshell that is slowly but surely preparing in this home of Africa, I don't believe they would sleep easy in their beds, even in London. Each rifle that finds its way in, in this casual way, will cost us English people at least 100 pounds, besides the lives of a good many of our fellow countrymen, some day or other." Djibouti served as the chief source for all these except a little on the part of the Italians.[21]

Guignony combined business with his official duties. He arranged to supply Menilek with 30,000 rifles at a very cheap rate, on condition that Menilek granted him a pass for 60,000, the balance to be sold by Guignony for his own benefit. Guignony sold the balance indiscriminately at the rate of £2 per rifle, and profited greatly on the whole transaction. According to an Ethiopian regulation, no rifles could be sold without the permission of the governor of Dire Dawa, and no more than ten could be sold to any one person. However, as Ward explained, "One can drive through any Abyssinian regulation with the almighty dollar, and they are generally framed so as to provide a source of income to the framer. Menilek arms his soldiers, the chiefs theirs, the sub-chiefs theirs." The saturation of arms had led to ammunition being sold in the open market in Harrar. Bullets served as coinage, ten to twelve cartridges going to the dollar.[22]

The British again pressed the French. In Paris, British ambassador Sir Francis Bertie called on Georges Clemenceau on August 4, 1908, regarding the Ethiopian arms trade. He said that England had previously led in the sale of arms to the natives of Africa, regardless of the consequences. The trade had now passed into the hands of the French. Bertie appreciated the difficulty that the French government might feel in restraining the traffic in arms, given the business interests involved, but he cautioned that the indiscriminate introduction of arms into Africa would lead to serious disaster for France. Clemenceau expressed astonishment at the information about the French consul at Dire Dawa being invovled in the arms trade.[23]

Pressured by the British and Italians to take some kind of action for appearance's sake, if nothing else, the French government did prohibit the import of arms on March 15, 1909. Merchant protests softened the government's position and yielded a one-year grace period, until March 15, 1910. Meanwhile, the French had delivered from Djibouti to Ethiopia 20,000 Lebel rifles and 2 million rounds.[24] French Foreign Minister Pichon wanted compensation if France adopted measures against arms traffic at Djibouti, prompting Bertie to muse, "What kind of entente cordiale is this?" As Grey remarked, "the fulfillment, by a friendly power with special engagements towards England, of a moral obligation, recognized and accepted as such by other powers, can hardly be regarded as presenting a basis for a claim to compensation."[25] In November 1909, the Italians inquired about rumors that French houses were selling 20,000 rifles to Ethiopia. Pichon said the sale had been made by the French government. Italy warned that Ethiopia would now sell their more antiquated weapons to neighboring tribes. The restrictions imposed on the passage of firearms through Djibouti remained farcical.[26]

The Ethiopian arms trade gained a new player when the Japanese entered the scene in 1910. In defeating the Russians in 1905, the Japanese had reaped a bonanza of Russian rifles. The Japanese government eagerly contracted with French agents to sell the surplus arms. By early December these rifles were on the way to the Red Sea and Djibouti. Count Komura Jutaro, Japan's minister of foreign affairs, said the shipments in question were in the ordinary course of trade, and added that if the Japanese did not ship arms the subjects of other countries would. The Ethiopian government had several Japanese rifle contracts under discussion. Guignony had offered Ras Tessama 20,000 Russian rifles that had been taken in the Japanese war. The regent wanted only 5,000, and Guignony proposed to sell the remaining 15,000 for trade purposes. This deal collapsed, but a few months later the Ethopian government was dealing for the same rifles with another Frenchman, Armand Savouré. A contract was made to supply 20,000 at once and 36,000 at a later time. The Ethiopian government paid for the first consignment in gold valued at 230,000 francs, which was to be forwarded to Paris through the Bank of Abyssinia. The French legation, however, supported Guignony and raised obstacles in the way of Savouré, and he lost a considerable portion of the expected profit. Soon the Ethiopian government was persuaded that it had made a bad contract and wanted to cancel. In 1911 the Japanese shipped 3,820 Russian rifles from Yokohama for Djibouti. These were followed by an additional 16,000 Russian rifles. An agent at Tokyo, on behalf of the Bank of Abyssinia, had contracted with Japanese authorities for the purchase, at 3 yen (6 shillings) apiece, of 60,350 rifles captured from Russia. The Japanese had already delivered 20,000. The *Kawachi Maru* sailed for Djibouti with 20 million rounds of ammunition in June 1911. This contract failed, and Japan held on to 40,350 Russian rifles.[27]

The Germans, too, now competed for a piece of the Ethiopian arms trade. At first the German legation offered to sell rifles from disused military stock for $5 apiece. Ethiopia declined the offer in December 1910. Early in 1911 the Germans tried again. Through an Austrian named Schwimmer and the German charge d'affaires Dr. Zechlin, the Germans proposed the sale of 30,000 German rifles left over from the Franco-Prussian War at $5 each. Next the Germans proposed to sell new weapons and modernize Ethiopia's armaments. The German legation worked to secure this trade by providing arms uniformity. The Ethiopians ordered 30,000 Mausers, and the Germans contracted for 250,000–500,000 more to unify Ethiopian armaments.[28]

A wholesale rearmament in Ethiopia raised the problem of what to do with all the old weapons. It seemed very likely that a massive flood of old rifles would inundate the region. Count d'Apchier, the new French representative in Addis Ababa, calculated that from 1901 to 1911 as many as 7 million rifles had passed through Djibouti. As a solution, British officer Charles Doughty-Wylie proposed a rearmament and disarmament idea on December 12, 1911, that would abolish private agents entirely. Instead, the Italian, British, and French legations in Addis Ababa would handle Ethiopia's orders cooperatively. Since Ethiopia lacked the money to pay for new weapons, Doughty-Wylie suggested that Ethiopia hand over six old rifles for every new rifle. The lure of new rifles would serve as an incentive for Ethiopia to seize all other rifles. Wilfred Thesiger wrote to Grey, "In this way they would have a strong inducement to disarm the frontier tribes first, which is what we most desire."[29] The obvious way would be to arrange for France to supply the new pattern of rifle. Unification of arms with a French pattern would make smuggling into British and Italian lands difficult because of the different systems.[30]

British officials advocated for the plan. Thesiger, the head of the British legation, heartily approved of the plan. To pay for its implementation, he suggested that in return for Ethiopian agreement to the proposed conditions, Britain should offer them £10,000 a year for five years. British colonial governments in Sudan, Uganda, East Africa, and Somaliland could divide this cost evenly. In anticipation that Ethiopia would require 500,000 rifles for rearmament, Thesiger calculated that "2,000,000 old rifles would be accounted for and taken off the market at a cost of 6d. apiece."[31] Doughty-Wylie explained to the Foreign Office that Britain could do nothing to prevent the rearmament of Ethiopia, and unless Britian implemented some scheme like Thesiger's, the Ethiopians would fall back on German assistance and input hundreds of thousands of new rifles, with the majority of current weapons ending up in the hands of frontier tribes. Grey endorsed the idea and passed it on to the Colonial Office. To convince the Ethiopians to go along, in a meeting of the three legations on June 27, 1912, Thesiger emphasized how the Ethiopian government, through agents, had received bad material at exorbitant prices. If the Ethiopians employed the legations as their sole agent, they could still place their orders in any country of the world, receive the quality of rifle they ordered, and would have to pay a commission of only 2 or 3 percent, thus saving the 40 or 50 percent profit pocketed by the current agents. What the three legations should have insisted on was ending the practice of private merchants being allowed to sell rifles.[32]

Paris and Addis Ababa began to exert pressure on French officials in Djibouti. Raymond Poincaré instructed colonial authorities to redouble their vigilance to prevent evasion of the embargo on imported arms for Abyssinia. However, cases of ammunition directed to the Djibouti government were forwarded to Guignony and sold by him at Dire Dawa and Harrar. Pascal, the governor in Djibouti, credited the profits from these transactions on the books of the French colonial government. The Italian consul at Aden visited Djibouti and determined that Pascal acted as the principal dealer in arms in the colony. Count d'Apechier, the head of the French legation in Addis, had a strained relationship with Pascal over the arms trade.[33]

France was faced with the potential entry of German arms undermining its effective monopoly. In 1912, it looked as if the French had decided to cooperate with the British and Italians. French cooperation held out the promise of genuine progress in controlling the arms flow into Ethiopia, and the tripartite agreement became, for the first time, something more than a piece of paper. The British credited Count d'Apchier for his assistance in temporarily closing Djibouti to all arms and ammunition destined for Ethiopia. In February the three legations met and formally protested jointly the abuse of the privileges secured to Abyssinia by the Brussels act. Closing Djibouti initially had little effect on the contraband trade, as the supply of rifles and cartridges for the general market did not diminish appreciably, and prices did not increase. While d'Apchier was trying to assist the British and Italian legations, in April–December 1912, the French permitted the official import of 75,000 rifles and 49 million cartridges. Because of this, the legations put forward a plan to rearm the Ethiopian army with one type of rifle. The Ethiopians had been discussing this with the Germans. The three legations proposed that Ethiopia tender for new rifles in Europe through them, thereby cutting out the private traders. The new rifles would be delivered on receipt of four old rifles for each new one.[34]

The proposed disarmament scheme put forward by the British legation never materialized, failing for internal and external reasons. The British Colonial Office deemed it too expensive, and therefore the British never pursued the buy-up program. Then Menilek died, causing political confusion and uncertainty for the Ethiopian government. With Menilek gone, his former subordinates moved to secure their own supplies of arms. Ras Michael, acting on his own behalf, negotiated at Addis Ababa to purchase 60,000 Turkish rifles from the Greek government that had been captured by the Greeks in the Balkan War. Other Ethiopians approached the French to place an order for 10,000 Gras rifles and 1,000 cases of cartridges. Competoir de Jibuti, Kevorkoff, and Garigue ordered 1,000 rifles and 10,000 cases of ammunition. In the absence of a strong Ethiopian ruler, the British now found no political partner with whom to work for the suppression of the arms traffic. Meanwhile, the closure of Djibouti had increased the smuggling by dhows. Consequently, by mid-1913, French authorities had eased up on restrictions.[35]

The French returned to their arms trade with renewed vigor. In 1913 the Ethiopian government gave its final consent to the construction of the last section of a railway to Addis Ababa by the French. Ethiopia signed the Vitalien railway concession on condition that the French remove the embargo at Djibouti regarding the transport of arms. An agreement was signed in May, by which the company agreed to pay Ethiopia 10 percent of the cost of construction (estimated 3 million Maria Theresa dollars). Arms and ammunition began arriving from Djibouti by rail in August. With the ease of transport provided by the railroad, at least 1,000 rifles a week entered Ethiopia.[36]

As the emperor weakened physically after 1910, so did Ethiopian state centralization. As a result, the unceasing flow of the arms trade contributed to increasing destabilization and violence. Exasperated by the French, the British contemplated one other option. As long as Djibouti remained French, the British could expect no improvement in the situation. Therefore, Thesiger mused about either purchasing Djibouti from the French or trading them Gambia or a territory in another part of Africa for it. Such an exchange could destroy the arms traffic at its source, and cut off the supply from which the mullah obtained rifles and ammunition and threaten British colonial borders. In the question of ultimate partition of Ethiopia, the removal of the French would strengthen Britain's position for the inevitable breakup of Ethiopia.[37]

Given the recognition of Ethiopian independence and sovereignty following the defeat of the Italians, the arms trade after 1896 no longer served the same purpose it had in the earlier period. By the early twentieth century Ethiopia was awash in arms, not only for national defense. Prestige, patronage, control over munitions, and trade interests all played a part. The Anglo-French entente formed in 1904, but did not have an appreciable effect on reducing friction between the two powers when it came to the French arms trade into Ethiopia. The French in Djibouti became dependent on receipts from the arms trade to pay for colonial administration. Contrary to British fears, the French arms trade was less a conscious challenge to British colonial control in East Africa than a fiscal source for Djibouti and a lucrative source of private income for French merchants and colonial officials. For revenue reasons the volume of the French arms trade escalated after 1898, and Djibouti transformed into a colony whose whole existence revolved around the arms trade.

4

Austro-German Hegemony
in Eastern Europe

The buyer states of Eastern Europe in the years 1880–1905 present a useful range of case studies to explore the interactions of diplomacy, business influence, and political systems. The Ottoman and Russian empires remained autocracies, yet they determined their arms procurements very differently. As autocrat of the Ottoman Empire, Sultan Abdulhamid II (1878–1908) routinely involved himself personally in the arms procurement process. Rampant bribery and the sultan's whim often overruled the professional opinions of his armed forces in selecting armaments. Moreover, the sultan overtly tried to gain German prestige and diplomatic support through his arms purchases, in the process giving the Germans virtual hegemony in the Ottoman arms trade. Tsarist Russia demonstrated a growing specialization and professionalism within its bureaucracy and armed forces. The tsars did not regularly intervene in the procurement decisions. Instead, they left the task to professional military and naval commissions to conduct equipment trials, to evaluate performance, and to deliberate prices. Matters of technical quality, performance, and cost generally shaped tsarist arms purchases as the Russians sought to build their own domestic suppliers as much as possible. Although Russia entered a formal military alliance with France, French firms did not achieve a lock on the Russian arms market.

In the aftermath of the Russo-Turkish war of 1877–1878, Romania and Serbia joined Greece as independent kingdoms. Bulgaria received autonomous status as a principality within the Ottoman Empire. The four Balkan states functioned as constitutional monarchies with elected parliaments. The four states' prime objective was national territorial expansion, to claim what each deemed as ethnic or historic boundaries. The conflicting territorial aims resulted in national defense with an eye to offensive expansion as Bulgaria, Serbia, and Greece each coveted Ottoman-controlled Macedonia. Additionally, the Serbian rivalry with Bulgaria erupted into war in 1885. After active involvement in the arms trade to the Balkans prior to the Russo-Turkish war, Russia had very little to show for it by 1885. In terms of relations with the Great Powers, by the mid-1880s Serbia had concluded a number of agreements with Austria; Bulgaria had moved out of its assumed Russian orbit into a pro-Austrian position; and Romania had signed a defensive alliance with Austria aimed against Russia. Austria and Germany had formed the Dual Alliance in 1879, and as the result the Balkan states became diplomatically tied to Berlin and Vienna either directly or indirectly.

Generally, Eastern Europe's place as the principal buyers' market in the international arms trade decreased in importance beginning in the late 1880s. Nevertheless, the region continued to play a crucial role in sustaining new armaments firms from 1880 to 1905. Where U.S. rifle firms had recently floundered, German and Austrian firms prepared to take the field. Mauserwerke, the German rifle factory at Obendorf, became a global arms supplier with sales to Eastern Europe. This success began when Wilhelm Mauser, facing dim prospects for future work and imminent bankruptcy in 1879, set off for Belgrade. The Serbian government effectively delivered his firm from pending collapse by ordering more than 100,000 rifles the following year.[1]

The main Austrian rifle producer, Österreichische Waffenfabriks-Aktiengesellschaft in Steyr, also found salvation through exports to Eastern Europe. For Steyr, domestic sales for its Mannlicher rifles evaporated from 1879 to 1886, causing its workforce to decrease from 6,000 to 910 workers by 1884. Only the company's Balkan sales, through Romana, carried the factory through those hungry years. In 1879 Romania decided to modernize its Peabody rifle system, and Bucharest turned to Steyr to furnish 130,000 rifles into the early 1880s.[2]

The German company Schichau, located in Elbing, owed its rapid rise as a major producer of torpedo boats to the appetite of an Eastern European customer. The Russian navy conducted tests of the boat in St. Petersburg on November 28, 1877. Subsequently, the German and Russian governments placed orders, but the Russian order proved larger: Russia bought ten Schichau boats compared to Germany's meager purchase of two. From this quick start Schichau would turn into the premier foreign supplier of torpedo boats for Russia. From 1877 to 1886 Schichau sold sixty boats domestically and exported forty-three, of which Russia bought twenty.[3]

By 1880 the Ottoman government faced bankruptcy, and the large Turkish war indemnity to Russia threatened catastrophe. To stave off fiscal collapse Sultan Abdulhamid II issued the Decree of Muharrem in 1881, creating the Ottoman Public Debt Administration (PDA). The Ottomans derived considerable benefit from the PDA, as half their debt was forgiven and borrowing from European sources became easier. On a negative side, though, was the PDA's authority to collect its own taxes within the empire. In this way much of the state revenues that could have flowed into the government's hands went instead to the PDA. From 1854 to 1914 Ottoman gross borrowing totaled TL399.5 million, and 45 percent of the loans were used to liquidate debts. An additional 34 percent was taken out as part of the commissioning of the loans, and only 6 percent (TL22.3 million) was spent on the military. The sultan had saved the state from financial ruin, but at the expense of not paying teachers' salaries or buying the technological means for the empire to defend itself based on its own domestic resources.[4]

The period 1885–1895 was marked by the rise of German defense imports to a position of virtual hegemony, and the Germans owed their newly found advantageous position to the policies and temperament of the sultan, who was highly autocratic, deeply suspicious, and overly controlling. The Austrian ambassador, Baron Giesl von Gieslingen, described the sultan's governing system as based on the pillars of "centralization, despotism, and corruption."[5] The sultan had an obsessive preoccupation with loyalty and relied heavily on a personal network of internal spies and police informers. His personality thus promoted corruption and favoritism, as he valued personal loyalty over efficiency or performance. The effects of his autocratic style could be seen in the Ottoman armed forces. The sultan distrusted the Ottoman navy because he knew that it had played the key part in the coup that deposed Abdulaziz. Consequently, the new sultan treated the Ottoman navy as an unwanted stepchild. Ever fearful of conspiracy, Abdulhamid kept his forces on a short leash. The navy was forbidden to leave its docks on the Golden Horn, to forestall any potential movement against the palace, and the ships rusted away at anchor. The sultan was also wary of the army because he suspected that the military college promoted liberalism among its students. Therefore, he consciously promoted and favored officers who had not been trained at the academies and who lacked background in modern military science.[6]

The sultan sought a German Military Mission in the aftermath of the Turkish defeat by Russia. In June 1880 he requested that officers of the German General Staff, infantry, cavalry, and artillery services come to the Ottoman Empire on a three-year contract. In April 1882, officers Colonel Köhler, Captain Kamphoevener, Captain von Hobe, and Captain von Ristow arrived, and the sultan gave them ranks within the Ottoman army. Later that same year Colmar Freiherr von der Goltz joined the mission. After Köhler's death in 1885, von der Goltz functioned as acting head of the mission. Goltz-Pasha remained in the Ottoman realm until 1895, and after his departure the influence of the mission declined. By 1898 only three of the reformers were still in Turkey.[7]

Abdulhamid had his own reasons for demonstrating a preference for Berlin. In the years 1876–1877, prior to the Russo-Turkish war, the Ottoman War Ministry had engaged a considerable number of British officers. The Turkish political objective in employing them was the hope that their presence would be followed by active English intervention in favor of Turkey. After the Berlin Congress (1878), when the British had not only refused to stand up for Turkey but also had deprived it of Cyprus, the sultan made every effort to minimize the authority and influence of the British officers. Prospects for a French mission also seemed unlikely because the sultan distrusted France as much as Britain. Therefore, the sultan looked to Germany as the first military power in Europe and as the most disinterested regarding Turkey. Abdulhamid wanted to retain good offices at Berlin, and he hoped that in calling a German mission he could make the world believe that Turkish military power was being fortified with the best instructors and weapons.[8] However, his distrustful nature meant that no foreign military mission would ever be treated with real confidence and no foreign officer would ever be given serious exclusive authority, except under the exigencies of hostilities. Moreover, the foreign officers discovered to their dismay that "zeal and industry on their part are discouraged and are positively distasteful."[9] Indeed, von der Goltz complained repeatedly about his inability to effect more improvements in the Ottoman army. Meanwhile, according to Henry Woods, an English naval officer in Ottoman service, Ristow-Pasha, the artilleryman in the German mission, "found so little to do that he spent much of his time in Janni Bier Halle in Pera, and attained the distinction of being known as Beerah Pasha."[10] Specifically, the sultan forbade training maneuvers because he feared that they could lead to a military coup, and he never allowed the troops to practice with live rounds.[11]

The sultan requested and obtained from Berlin a German Military Mission to oversee training and modernization of the Ottoman army. For their part, the Germans sought to use their privileged position to support German military suppliers as they overhauled the Ottoman armed forces.[12] Because the mission was incorporated within the framework of the Ottoman military system, it had tremendous opportunities to expand the role of German arms and equipment in the Ottoman army. Primarily, this unique position allowed the German officers to have easy and frequent access to the Ottoman Ordnance Ministry. Since this ministry controlled supply and military production, the Germans now had a direct channel for shaping Ottoman armament policy. Until 1889 all the German officers were subject to the serasker (war minister), but early in 1889 von der Goltz declined to renew his contract because he was dissatisfied with his lack of influence. As result, he was attached directly to the Imperial Military Household with the right to address reports to the sultan. The German mission caused some grumbling among Turkish officers and foreign observers. As the British noted, "One of the main handles for intrigue has been the accusation against the Chief of the Mission being interested in supplies of war-like material by German firms . . . but in 1891–92 there was an active but unsuccessful intrigue by the then French Military Attaché in favor of a French Military Mission."[13]

The German Military Mission proved a boon for German firms. In 1882, the Ottoman government placed an order with Krupp for TL1,206,987 worth of artillery to replace and repair the fortifications and batteries of the Bosporus and Çanakale. This order was largely because of von der Goltz's insistence. In his memoirs, von der Goltz related how he tried to impress on Turkish military leaders and other officials that the independence of Turkish policy depended on the security of its capital, Constantinople. Accordingly, in 1885 he convinced the Turks to buy 500 Krupp heavy guns to defend the Dardanelles. The magnitude of this sale can be appreciated when one considers that in the 1877–1878 war the army in Europe had 590 field guns. Von der Goltz also noted the need for torpedo boats to augment the defense of the straits, and he pointed the Turks to Schichau to provide the necessary wares. In 1886 the Ottomans bought 426 field guns and 60 mortars from Krupp, and ordered torpedo boats from the German firm Schichauwerft. In reflecting on the orders of the 1880s, von der Goltz proudly proclaimed his special service to the Fatherland in rearming Turkish forces with weapons from Germany where previously they had been supplied by England and France.[14]

Undoubtedly the Germans profited from these sales, but von der Goltz gave himself and the German mission too much credit. In actuality Abdulhamid II acted as the decisive force in granting the orders to Germany by playing a direct role in making the armaments selections. For example, in 1887 an Ottoman Military Commission was deliberating between the Mauser or the Martini as the new rifle system. The previous year, German rifle manufacturers Paul Mauser and Isidor Loewe had met May 2, 1886, and decided to cooperate in securing the Turkish order. Subsequently, Mauser traveled to Constantinople to persuade the Turks to buy a variation of their rifle model 1871/84. On November 17, 1886, Mauser met with Abdulhamid, and on December 8 the Turks began field tests of two different Mauser models (11mm and 9.5mm) along with Belgian, Austrian, and British Martini and Hotchkiss rifles.[15] Over the objections of the Ottoman War Ministry and other Turkish military authorities, the sultan ordered a provisional contract with Mauser. Only financial difficulties prevented the provisional contract from being finalized.[16] In Constantinople agitation against the Mauser contract was growing, but that did not matter since the sultan considered the deal "as his own act and deed."[17]

In the end the sultan's will prevailed. In February 1887, the efforts of Mauser and Loewe were rewarded when the sultan signed a contract that divided the order evenly between their firms. The unit price was 362 piasters (68.8 marks) per rifle. Besides ordering 500,000 rifles, the Ottoman government also purchased 50,000 carbines. In placing this order the Turks became the first army to acquire the Mauser magazine rifle in any significant number. Fearing that they would be left behind by ongoing technological developments, the Turks stipulated in the contract that should a better rifle be produced during the fulfillment of the order, any undelivered rifles would be upgraded to the latest model. The Turks actually invoked this clause when they learned of the model 1890 rifle, which used smokeless powder instead of black powder and required a smaller caliber round. As a result, the Turks obtained a majority of the newer model (280,000). Yet by 1890 none of the Mausers had been issued to the troops or training schools; instead, they remained uncrated in storage.[18]

In the years ahead, the sultan continued to make large arms purchases from the Germans and resorted to foreign loans to pay for them. In 1889 the kaiser visited Constantinople, and the sultan showered massive new war orders on the German firms. Krupp sold artillery, Mauser and Loewe sold rifles, and Schichau sold torpedo boats for a total of 15.3 million marks. Such a large order required a loan to finance it, so Deutsche Bank acted as contractor, accepting the income from the PDA's fishing industry as security for the loan value of TL1,617,647. When put into context with the annual expenditures in the Ottoman budget, the enormity of this purchase becomes apparent. The loan represented 19 percent of the total military-naval budget and about 10 percent of the total Ottoman revenues for the year. Between 1891 and 1894 the Ottomans bought an additional 35.1 million marks' worth of armaments. Having committed to the Mauser system in such large numbers, the Ottoman government remained one of Mauser's best customers. In 1893 the Turks ordered an additional 201,000 new rifles based on the Spanish model of that year (7.65mm). The Ottoman Treasury was weak, so the Turks had to take another loan. The nominal amount of the loan, arranged in Germany, was TL1 million, bearing 4 percent interest. The first installment of TL300,000 represented half the total purchase price.[19]

Friedrich Alfred Krupp understood that the key to winning orders in Turkey was the sultan. Abdulhamid II was the sole maker of policy, and officials at Sublime Porte were just mouthpieces. Even with these stunning successes, Krupp always worried about losing his position. He complained in a letter sent from Constantinople in late December 1891 that German influence was dwindling as French influence was on the rise, and he blamed the German ambassador for the turn of affairs. He feared that Russia and France would influence the sultan through their schemes, baksheesh, and bribery of those in the sultan's circle, and through their threats. Krupp wrote, "While earlier Krupp, Vulcan, Germania, Mauser, and Rottweil supplied the guns, ships, rifles, and munitions of Turkey, now these orders are made in France . . . while earlier it was the hope that Turkey in case of war would attach itself to the Triple

Alliance, this prospect in my view has become very doubtful."[20] Krupp's representation of the state of affairs was grossly inaccurate, and most likely the letter was part of his recurring laments that the German diplomats never did enough to support his business. More incisively, only Krupp talked of the desirability of a Turkish alliance, whereas von der Goltz had talked only about getting Turkish orders for German businesses. In truth, the German position remained secure. Von Gieslingen complained on his arrival to Constantinople in 1894 that the Germans had a monopoly on war orders for Turkey and that German business interests held the palace camarilla in their hands. In particular, the Austrian ambassador found it extremely difficult to win favor for Austria-Hungary's factories in the struggle against Krupp and other German firms. Von Gieslingen acknowledged that part of the difficulty was that, without exception, the German firms delivered first-class war materials.[21]

The French did make moves to gain ground in the Turkish market, but they knew they faced an uphill battle. As the Schneider company dossier about the negotiations put it, "The prized position of German industry does not permit us to enter into it with the conditions of normal competition."[22] Specifically, Schneider was pursing orders for warships and artillery for the Bosporus forts. Abdulhamid had sent a Turkish technical mission to France in May 1894. Schneider knew that in response, Krupp's agent in Constantinople, Karl Menschausen, had paid a call on the sultan in the summer of 1894 to present a model of fort plate for coast defense and to renew offers to the Ottoman government regarding the fortifications of Constantinople. Lieutenant Colonel Corbin of France had proposed a counterplan for the defense to Zekki Pasha that incorporated a grand fortress for two 24cm cannons, but Zekki Pasha wanted heavier guns—35.5cm. However, in June 1895 the new grand vizier, Said Pasha, seemed more favorable to French industry, and Corbin's plan was submitted to the palace. Said Pasha also expressed interest in acquiring an armored cruiser to French Ambassador Vicomte de Mareuil. When the ambassador pressed the vizier about the proposed fortifications, however, Said was evasive. The vizier did ask Mareuil to raise the financing question with the Ottoman Debt Council. In reply, Mareuil voiced his opinion that the Ottoman Debt Council might decide in principle to dispose TL3 million, some of which could be used to purchase armaments.

Meanwhile, Dreysse Pasha, an old French officer in Turkish service, had promised Mareuil to bring up Corbin's fortress proposal with the sultan at an opportune time.[23] Things appeared to be moving the French way. Mareuil reported to the Schneider firm that the situation regarding the fortress question had not changed. The political preoccupations and financial penury continued to cause delay. Nevertheless, after two years of efforts the French had created a project less costly and more practical than the German one, and successive Ottoman commissions had submitted it to the sultan.[24] Mareuil was optimistic that the French stood poised to take a prestigious Turkish order away from the Germans. To provoke the final approval of Corbin's plan, the French firms of Schneider, St. Chamond, and Chatillon had contributed 13,000 francs for the budget of the Turkish mission in 1896, and the firms were prepared to pay as much as 25,000 francs.[25] Unfortunately for French ambitions, the plans came to naught. Although Creusot and St. Chamond had submitted bids TL22,000 below Krupp's tender of TL121,000 for a pair of 30.5cm fortress guns, the Germans retained the fortress armaments contract.[26]

The Germans also improved their arms trade position by the way they conducted the military reforms. For example, in 1893 von der Goltz declared that a number of Ottoman officers must become proficient in the handling and use of the newly developed German rifles. Furthermore, he stated that the munitions factory at Tophane would have to produce "dummy" cartridges, and that these new rifles should be given to the troops. Clearly, the close German ties with the Tophane-i Amire Nezareti were paying off. During this time the French and British were practically eliminated from the Ottoman arms market. The complete market sovereignty of Krupp and Mauser by 1894 led to their exploiting their advantage, and to the accusation from foreign observers that they were selling expensive and low-quality goods. In 1895 an additional 12.2 marks worth of goods came from Germany. All in all, in the decade 1885–1895, 100 million francs (£4 million) worth of orders for war material went to German enterprises.[27]

Vincent Caillard, a board member of the Ottoman PDA, presented a report on the financial state of the Turkish empire in 1896. Caillard could not determine the exact amount of the regular annual deficit, because the irregularities of the autocratic government omitted many expenses from the budget. Nevertheless he estimated the annual deficit between TL800,000 and TL1.2 million. He reported:

The deficit is practically wiped off by the simple method of the Government not paying its officials and its soldiers, and leaving such creditors as it can unpaid . . . the total extraordinary resources in cash which have been placed at the disposal of the Turkish Government between 1890–96 amount to TL7,062,629. From this it is necessary to deduct an extraordinary expenditure upon the rearmament of the military forces with Mauser rifles and Krupp guns, amounting roughly to TL3,300,000, leaving a net cash amount of, in round figures, TL3,760,000.[28]

Based on Caillard's figures the sultan had spent almost half the funds on imported armaments.

It is difficult to gauge domestic production of military supplies, rifles, and other weapons for the army in the Hamidian period. Clearly the Turks were having trouble paying for plant maintenance and foreign experts. During the 1880s seven British workmen at Tophane filed petitions over wage arrears amounting to TL6,000. The claims occupied the attention of the British embassy from 1882 to 1889, but owing to the "impecunious state of the Ottoman Treasury" no settlement was forthcoming. Indeed, since April 1888 the Ottoman Bank had refused to pay anyone's salary, including the German generals or the Ottoman diplomatic service.[29] Eventually the British employees at Tophane did receive back pay for May–September 1888, but claims for arrears from the previous five years remained outstanding.[30]

By and large the equipping of the Ottoman army became the province of foreign imports. The Germans were especially important in this regard. During the 1890s, the mainstay of the Ottoman infantry became the Mauser rifles. According to Russian military intelligence, by 1902 the Turks possessed approximately 900,000 Mausers (220,000 Mausers model 1887, 280,000 Mausers model 1890, and 400,000 various other Mausers).[31] Here again, the Ottomans found it easier to buy the newer models as they came out.

The late 1880s saw little new naval construction, and the domestic naval program of the latter part of Abdulhamid's reign proved to be largely stagnant. In 1900 Captain Hugh Williams of the Royal Navy reported to British Admiralty Intelligence that for all practical purposes the Ottoman fleet had ceased to exist. He observed:

The larger vessels building in the Golden Horn are placed in impossible situation, and on badly prepared slips in the open, while there is a very fine covered building slip empty; it is therefore a fair deduction to assume that these attempts at building large modern vessels are unreal and only to show those outside the dockyard that those inside are really doing something, and that of a very up-to-date nature . . . The reason for this really sad state of affairs is not quite on the surface, for I believe a fairly large sum is devoted to the navy annually, but as a Turkish naval officer said to me: "To understand the condition of our service you must be acquainted with the present history of the Palace," which I believe to mean that, first of all, the navy is not trusted, and therefore the ships are kept at Chanak.[32]

The years 1885–1895 also marked a change in the pattern of Ottoman naval purchases as the Germans came to dominate Ottoman naval orders. Expediency, cost, and Abdulhamid's personal involvement all played a part in bringing about the turning point. Initially the Turks had hoped to buy more boats from France. But the French informed the Turkish Admiralty that no torpedo boats would be available for immediate purchase, so the Turks solicited bids for the boats. Shortly after the delivery of five boats, Schichau submitted a proposal offering eight vessels for TL257,300. As soon as other German manufacturers got wind of Schichau's offers, they entered the fray. Howaldt, Vulcan, and Germania submitted their own proposals. Germania offered the same boats as Schichau but undercut the latter's price at TL190,000. The French firm Forges et Chantiers was prepared to tender at a reasonable price, but declared that Germania's stated price was impossible once it factored in the baksheesh demanded from the minister of marine on down. Admiral Livonius, the chairman of Germania, visited Turkey, and after long-drawn-out negotiations did receive the contract, not only for the eight boats at TL190,000 but also for four additional thirty-nine-meter boats at TL16,000 each for a total contract of TL254,000. However, the Germania matter did not end there. The sultan granted his approval, but the naval department had to draw up the contract, and it put in some very stringent clauses. In particular, the contract bound Germania to guarantee security for repayment of all advances if the firm failed to deliver all the vessels at the low contract price. The final deal called for twelve boats at a lower price than the French had proposed for eight boats.[33]

In the ongoing competition with Armstrong, Krupp gained the upper hand. From 1886 to 1890 all five Sinub-class wooden sloops were rearmed with Krupps. In 1891, many of the Ottoman ships that had previously mounted British Armstrongs had them replaced with Krupps. Additionally, the torpedo gunboats and third-class cruisers built at Constantinople possessed Krupps. In terms of ship production, German firms gained substantially. No German-built ships had existed in the Ottoman navy in 1877. In this new period the German firms produced thirteen torpedo boats. The French did sell six torpedo boats to the Ottomans in 1885–1886 (La Seyne and Des Vignes built three each).[34] However, there were no more French orders after 1886. Therefore, 1886 can be seen as the beginning of German hegemony in the Ottoman naval market.

Obviously, British firms lost the most to the Germans. The true magnitude of German gain showed in the almost complete removal of British suppliers from a naval market that they had commanded previously. Britain still had a minor market in supplying Whitehead torpedoes, but this was a mere fraction of its former business. The sultan did like the English-made Nordenfelt submarine and personally purchased a pair. This act seemed more a vanity, as "His Majesty has already paid, not only the value of the material obtained from England, but the cost of putting them together here, the latter operation having taken three times as long and cost three times as much as the estimate."[35]

In the late 1890s the Ottoman government embarked on a foreign building and reconstruction program to reorganize the fleet following its poor showing in the 1897 war with Greece. In response to Turkish solicitations Thames Ironworks of England proposed to assume administration of the Turkish docks in Constantinople on a five-year lease and to modernize the facilities. As part of the terms, Thames Ironworks would build three 10,000-ton battleships and three 6,000-ton cruisers using materials imported from Britain. Abdulhamid declined the offer, purportedly because of Russian objections. Soon after, the Turks began discussions with Krupp and Schichau. Turkish insiders leaked the details to Armstrong, supposedly because Armstrong had been promised a Turkish contract as compensation for Krupp artillery orders. It is also possible that the Turks' motive was to drive down the cost by making Armstrong privy to Krupp's tendered price. This was, in fact, what occurred since Armstrong submitted a bid of TL2 million, 30 percent below Krupp's price.

Not wanting German prestige to suffer by letting the Turkish business slip into British hands, in October 1897 Kaiser Wilhelm II urged Krupp and Schichau to rebid and also to bring Vulcan into the German consortium. Friedrich Krupp reported to German Admiral Gustav von Senden-Bibran that the firms of Krupp and Schichau, unbeknownst to Vulcan, had already struck a deal in August regarding Turkish orders. The plan was to handle the Turkish business jointly and thereby foreclose any bilateral competition. Krupp insisted that if Vulcan chose to join the consortium, Krupp's delegate would remain the directing force. By December 1897 a modified Turkish program for rebuilding two old vessels and building eight new ships (two battleships, two armored cruisers, two protected cruisers, and two light cruisers) met with Abdulhamid's approval. At this stage Krupp and the Germans withdrew, saying that Turkish plans were unfeasible at the anticipated prices.[36]

Following the German withdrawal, the sultan took the opportunity to use his naval contracts to solve some diplomatic disputes. In 1898, the contract for rebuilding the ironclad *Mesudiye* went to Armstrong-Ansaldo, a branch of the British firm in Genoa. By giving the order to the Italian plant, the sultan hoped to satisfy Italian claims for damages arising from property destroyed during the Armenian massacres in 1895–1896. After initiating informal discussions with the Italian ambassador in Constantinople concerning Ansaldo's capabilities, Abdulhamid decided to involve himself personally in the refitting of the *Mesudiye*. On June 13 the sultan granted the contract to Ansaldo for TL433,560 to complete the ship's transformation over eighteen months. In June 1900, the Ottoman government signed an agreement with Ansaldo for four corvettes and four frigates valued at TL552,000. Then, in 1902 the Turks negotiated with the Italian firm to modernize their naval arsenal by creating a filial works in Constantinople, although this project was abandoned in 1906. The U.S. government had also submitted claims for damages from the Armenian massacres, but the sultan initially rejected those. When the Americans threatened to dispatch a squadron to Constantinople, Abdulhamid decided to compensate the United States by placing an order for one of the new cruisers with a U.S. firm. Cramps Shipbuilding of Philadelphia won a Turkish contract worth TL355,000 to build the protected cruiser *Mecidiye* in May 1900. Cramps delivered the warship in December 1903, but the Turks considered the ship a disappointment. The second Ottoman vessel to be rebuilt, *Asari-Tevfik*, went to Kiel for modernization by Germania in March 1900, based on the firm's price of TL280,000.[37]

The German dominance in the Ottoman naval market proved short-lived as the British firm Armstrong made a rousing comeback. Armstrong had first hoped to build a protected cruiser for the Turks back in June 1898 when the Turkish envoy Ahmed Pasha had met with Sir Andrew Noble in Newcastle. In 1899 Armstrong had tendered for a 9,000-ton battleship but nothing had come of it. Two years later the tender had been revived and Armstrong's board had reason to believe the company would receive an order shortly. By September 1901 the Turks had made a contract with Armstrong to build the *Abdulhamid,* but the Ottoman government did not pay the first installment, and no work was started. Finally, in October 1902 Armstrong received a payment of £33,000 on the £456,000 owed. The Turks also had arranged to make monthly payments of £30,000 from the Ottoman Bank. Armstrong laid the keel in November, and the Turks had it in hand on May 29.[38]

The sultan's motive behind the German arms purchases was less about improving the strength of the Ottoman army and navy and more about winning favor from Berlin. The sultan, the firms, and the German mission all wanted the sales, so it would be incorrect to attribute the sales primarily to the influence of the German Military Mission and von der Goltz. Krupp and Mauser understood that it was the sultan who mattered in arms sales, and they dealt with him personally. Von der Goltz also put his complete support behind German sales, but Abdulhamid probably would have bought from the German firms in any event. Krupp had already established itself in the Ottoman Empire in the previous era and had supplied superior equipment for the Turks. American rifle companies had gone bankrupt due to Turkish default in 1880, and only the German firms had been willing to take the risk of Turkish contracts, thanks to payment with Deutsche Bank loans. After 1897 the sultan used naval sales as a way to resolve other outstanding issues, such as compensation for Armenian damages, instead of using sales for naval modernization. Expensive warships rusting away and thousands of Mauser rifles still in boxes revealed the sultan's preference for the façade of armed strength without the risk that well-equipped Ottoman troops and sailors might turn on him.

During the 1880s Russia's role as a buyer and seller in the armaments trade changed. As Russian domestic production gained ground, the tsarist government weaned itself off imported Krupp artillery in preference for Russian-made guns. The Franco-Russian Military Convention of 1891, followed by the formal Franco-Russian alliance against Germany in 1894, brought a partner eager to beef up Russia's military and industrial strength as quickly as possible. A French government-owned plant, in cooperation with Russian state factories, successfully left Russia in possession of an updated rifle industry in the 1890s. Yet during the Russo-Japanese War, Russia purchased ammunition from German and Austrian firms. Moreover, the alliance did not give French naval sales an exclusive position in the Russian market, nor did it preclude German or British defense business. The Russians employed British and German models, and Russia bought more Schichau boats than any other country. Russia refrained from acting as an arms supplier to the Balkans, leaving the field clear for Austrian and German firms.

When Russia embarked on small arms modernization in the 1890s, the government followed the same strategy as in the preceding era. Having settled on the Russian-designed Mosin-Nagant rifle as its new magazine-fed weapon in 1891, the tsarist government placed a large order abroad while it retooled its three plants at home. Accordingly, Russia ordered 500,000 Mosin-Nagant rifles to be manufactured in the French state-owned rifle plant at Châtellerault, and imported over 1,000 machine tools from France, England, and Switzerland to equip the rifle plants at Tula, Izhevsk, and Sestrorets. Ordering abroad saved time, since France had a production capacity of 1 million rifles annually compared to Russia's 250,000. By the completion of the production run in 1896 France had delivered 503,539 rifles and Russia had produced 1,470,470 combat rifles.[39]

To make Russia independent of other states for armaments, the tsarist government began to encourage domestic private enterprise. As a result, one of Krupp's best customers now favored a domestic challenger. In 1900 the Putilov Company of St. Petersburg made its move to establish itself as the chief supplier of field artillery to the tsarist government. In 1899, the Russian army announced that it would be conducting trials to select a new quick-fire system to rearm the artillery branch, and the Rearmament Commission would decide the contract based on quantity and price. The Main Artillery Administration (GAU) used its state plants as a cost control, and insisted on pricing consistent with its own estimates. Krupp, Schneider, and Putilov guns were all entered in the competition. The Putilov gun performed superbly, but Putilov did not receive its first order for 750 Russian-designed 3-inch field guns until after lowering its initial bid price at the suggestion of the GAU. The contract was signed in June 1900 for a cost of roughly 9 million rubles.[40]

During the war with Japan, Russia was unable to meet its supply needs from domestic suppliers. In May 1905 the Russian government ordered 300 million rounds of rifle ammunition from Germany and 200 million from Austria. The British firm Kynoch Company had bid for part of the order, but its offer of more than 50 rubles per 1,000 was too high. Kynoch did offer a lower price of 40 rubles, but by then the contract had already been signed. Political considerations also played a part in rejecting Kynoch, "owing to the direct veto of the Emperor out of a delicate regard for British neutrality."[41] In January 1905 Böhler Company in Austria obtained a Russian order for 500,000 rounds of shrapnel, 22 rubles per shell. Schneider also obtained an order for 500,000 shrapnel rounds for 20 to 21 rubles, but said it would take eighteen months to fill the order. Ehrhardt of Germany also received a comparable order for 500,000 at 17.75 rubles.[42]

Ehrhardt (Rheinmetall) offers yet another example of a German armaments firm saved by Russian contracts, in this case worth 20 million rubles. Ehrhardt's foreign contracts the previous fifteen years had resulted in losses on the company's balance sheet. After receiving orders from the tsarist government, by 1905 Rheinmetall finally turned a profit, began to pay off its bank debts, and reacquired a factory building that it had lost to Germania in 1901.[43]

After the Russo-Turkish war of 1877–1878, a Russian warship construction program called for the Baltic fleet to add sixteen battleships, thirteen cruisers, eleven gunboats, and one hundred torpedo boats. As the prototype for domestic construction of torpedo boats, Russia used the British Yarrow boat. Early in 1877, prior to the war, Captain Kasy of Baltic Shipbuilding had come to Yarrow's yard in Poplar to arrange for the English firm to sell designs for a torpedo boat small enough to be transported by rail from the Baltic Sea to the Black Sea. Yarrow provided the drawings for hull and machinery and also offered to supply engines and boilers from its plant. Kasy accepted the drawings, but the Russians turned to German firms for the machinery because the tsarist government wanted to avoid a delivery disruption from Britain in case of war with Turkey. At the same time Captain Gulaev, a member of the Russian Navy Technical Committee, purchased two larger Yarrow boats using funds raised by popular subscription in Russia. The Russian concerns about complications with Britain proved true, as after the outbreak of war with Turkey the British government prevented the departure of the two Yarrow boats bought for the Russian navy. In response to Russian demands that the money be returned, the British government stepped in and bought the boats. After the war, Russia went back to the previous plans for one hundred boats, and in 1883 the Baltic Shipbuilding Works began building the first ships of the new program.[44]

Russian satisfaction with the Yarrow designs brought the company repeat business. In 1880 the Russian Admiralty wanted larger torpedo boats for the Black Sea fleet. The Russians resolved to place an order with Yarrow for a sample boat, the *Batum*. It was faster than the other boats, and sailed to the Black Sea at twenty-two knots. In 1895 Russia again bought a fast Yarrow torpedo boat destroyer, the *Sokol*. This boat contained a number of technical advances. *Sokol* was the first destroyer capable of thirty knots, and its construction used a new class of steel of higher tensile strength. Once again, the Russians used a Yarrow boat as a model for domestic production, and they began building copies of the *Sokol* in Russian yards.[45]

The Russian navy had been a major purchaser of Krupp ordnance, but after adopting the Krupp gun for its armament, Russia began to manufacture that system for its own use in the 1890s. Until 1891 the Krupp system had been applied to all Russian naval guns above six-inch, but after 1891 all heavy guns would use the French Canet system for closing the breech. Canet's system of quick-fire guns for the heavier guns was also adopted. The Russian government concluded a contract with Canet's firm for drawings and the right to manufacture 75mm, 12cm, 14cm, and 15cm breech mechanism and mountings for 200,000 francs (£8,000). Russia also looked to purchase Armstrong drawings and manufacturing rights for quick-fire ammunition. The successful domestication of Krupp's techniques was evident in the declining costs of Russian manufactured pieces. In 1886 the relative costs of Russian-made eleven-inch Krupp guns made them cheaper than imported ones. Krupp sold Russia ten such guns for £16,512 each, whereas Obukhov produced the same gun for only £13,090 each. By 1900 newly installed Krupp plants were in operation at Obukhov and Izhora. Up to 1903 Krupp made Russia's capped projectiles, while Russian ones were manufactured at Perm and Obukhov.[46]

Russia's ongoing diplomatic rivalry with Britain during the second half of the nineteenth century did not inhibit Russian desire to acquire English naval technological know-how. In terms of naval construction, in 1894 British Admiralty intelligence reported, "Speaking generally, the machinery and tools are of English make, and in many cases the rough material is English also. Nearly all establishments have one or more British foremen, but at the same time most of the designs for guns, gun mountings, turrets, barbettes, ammunition, etc., are now of French origin, and the tendency at present seems to be to follow the French lead in all such matters. There is a great desire to make everything for the navy in Russia."[47]

Foreign-built ships comprised about one-quarter of the Russian navy prior to the Russo-Japanese War of 1904–1905. These foreign suppliers included firms from Britain, Belgium, France, and Sweden. Among the forty-eight torpedo boats added to the Russian Navy in 1880–1894, imported ships numbered eighteen. Schichau supplied half the foreign boats, while six were French and the final three were British-made. In addition to vessels purchased abroad, many non-Russian firms supplied engines and boilers for warships built in Russia. These foreign suppliers included Fairfield Company (one), Cockerill (one), Napier and Sons (four), Mawdsley and Field (five), Humpreys and Tennart (four), Hawthorne and Leslie (two), Baird (four), Motola Co. in Sweden (six), Mannesman (one), and Belleville et Cie (three). In the years 1895–1905, foreign yards produced two out of fifteen battleships (Cramps and La Seyne) and three out of six armored cruisers (two La Seyne and one Vickers). By 1901 Germania had delivered the first-class cruiser *Aksold* while two French firms gained new orders for destroyers. The Forges et Chantiers de le Méditerranée (Havre) was building three destroyers and Normand an additional two. In 1898 for the Pacific fleet, the Russian Naval Ministry placed an order for thirty torpedo boat destroyers with Cramps in Philadelphia. These boats were delivered directly to Vladivostok. In that same year, orders for torpedo boats and destroyers that had been given to a Belgian firm were canceled and given instead to Kiel yards.[48]

Overall, the tsarist government, through the War and Naval ministries, based its procurement choices on technical quality and price. The case of the unarmored cruiser *Boyarin* proved the exception. For the construction of this ship, Laird and Schichau had offered better tenders, but the Danish firm Burmeister and Wain received the contract at the insistence of the Empress Maria Feodorovna, "the Imperial Family being so content with the royal yacht *Standardt* which was built by the same firm."[49]

The predominance of political considerations over economic ones in Russian arms sales to Bulgaria became evident when relations ruptured over Bulgarian annexation of Eastern Rumelia in 1885. Russia not only withdrew tsarist officers from service in the Bulgarian army in protest but also withheld ammunition from Bulgaria, including 8 million cartridges slated for delivery. The Bulgarian authorities found themselves in a difficult situation. Should they seek an alternative supplier and completely change the weapons systems in their armed forces, or should they endeavor to smooth things over with Russia to maintain their Russian Berdan rifles? The likely alternative supplier was Austria. However, Austria had allowed Serbia access to armaments during the Serb-Bulgar war of 1885. In the summer of 1886, the Prince of Bulgaria expressed his lack of faith in Austria, which could have long ago stopped the present Serbian armaments had it wished to do so. In July 1888 the Bulgarian government entered into a contract for approximately 30,000 Berdan rifles for 58 francs each with an agent that was to buy directly from the Russian government. Although this price was higher than previous costs, the deal would mean continuity in Bulgarian equipment. Most of the items appeared to be of inferior quality, and were weapons rejected by Russian military authorities. The contractors had said they would be unable to make delivery on November 1. The Bulgarians became suspicious that the syndicate was acting in connivance with the Russian government or that the tsarist government, aware of the deal, would impede the delivery of the rifles. The contract had not yet been executed, and by August 9 the Bulgarians had cancelled the contract. Five million Berdan cartridges were already on the way to Sofia from Austria, and Bulgaria was being offered good terms for 100,000 Mannlicher rifles in Austria.[50]

Having experienced direct political pressure from Russia over armament imports, Bulgaria now had to contend with economic demands from Austria. With regard to the Mannlicher rifle, three Bulgarian military commissions had decided against its adoption. On technical grounds, the decision seemed strange in view of the superiority of the Mannlicher over the Berdan and the old Russian Krinka, which was the chief weapon of the Bulgarian Reserves. Furthermore, the Berdan could be procured only in Russia. Bulgarian military officers felt that the Austrians were trying to foist the Mannlicher on them. The Austrians insisted on the purchase of a certain number of Mannlicher rifles as one of the conditions attached to the very unpopular loan of the Länder Bank of Vienna for 10 million francs. Lacking another option, Bulgaria eventually accepted Austrian terms. By May 1891 approximately 60,000 Mannlicher magazine rifles had been delivered with 40 million cartridges, and 30,000 more rifles were delivered within the year.[51]

Increasingly Bulgaria turned to Austria and Germany in place of Russia. For artillery, the Bulgarians turned to Krupp. By 1892 Bulgarian artillery possessed more than 300 Krupp guns of the most recent models. The Austrian Mannlicher 8mm had become the standard infantry weapon, and 70,000 Mannlicher cartridges had been ordered from Roth in Vienna. Bulgaria was also considering ordering some quick-fire guns from Gruson. In 1893 Bulgarian military orders from Germany amounted to 4.327 million francs, or a third of all German exports to Bulgaria (12,060,000 francs).[52]

Despite its limited economic resources, the Bulgarian government consistently and willingly bore the burden of increased military expenditure. The War Ministry received generous treatment from all political parties in budgetary allocations. The military budget increased from 17,227,000 francs (£649,080) in 1886 to 23,918,121 francs (£956,724) in 1890. By 1892 the War Ministry budget claimed nearly £960,000, or roughly 30 percent of the total state budget. The 36,170,000 francs (£1,446,800, 1886–1892) already assigned to purchase armaments and material, together with later special credits, and the 750,000 francs (£30,000) to prepare defense positions, provided for the necessities.[53] As the British military attaché Colonel Chermside reported, "all political parties in the National Assembly appear agreed as to the necessity of supporting this burden in the self-interests of the young nationality, as apart from considerations of defense of their liberties, or resistance to aggression by jealous neighbors, all Bulgarians unite in very definite aspirations as to political expansion, and surpluses of actual over-estimated revenue arising from good crops, etc are preferentially assigned to the War Ministry."[54]

Just a few years later the War Ministry pressed for still more armaments. The Bulgarian war minister, Colonel Racho Petrov, addressed the Bulgarian legislature (sobranie) in February 1896. He highlighted the industrial weakness of Bulgaria and the need to develop some domestic production capability. He stated:

> If we were at war today it would be impossible for us to procure the necessary amount of projectiles and powder, for I cannot buy them in any baker's shop here. Consequently, my desire is to establish in the country a factory for arms and powder. But till then we must carefully amass supplies upon supplies. We must have rifles for 120 battalions of infantry, and 480 guns for the artillery, if we wish to mobilize only the standing army . . . Our old guns are no longer of any use; the artillery must be completely equipped.[55]

French, Austrian, German, and British firms all gained orders as part of the new rearmament program. Bulgaria became interested in the French firm Schneider when the Bulgarian minister of war, along with some officers, came to Bucharest to study fortification works. Impressed by the steel cupolas, they inquired whether a French group would come to Sofia. Reporting from Sofia in 1896, Maurice Michel-Schmidt calculated that support for the Franco-Russian alliance would possibly lead to an order of new war materials for French industry. Under such circumstances, the French minister estimated that a double effort by Creusot and Canet would be very favorable. The Bulgarian politician Nikolov had requested a study of cannon from Creusot, but Michel-Schmidt deemed Nikolov to be unpopular and not well thought of by the prince.[56]

Nevertheless, Schneider began to assert itself for a share of the Bulgarian market. As a key weapon in its arsenal, Schneider tried to deploy French financial strength. French finance intervened more directly in 1896 in negotiating a Bulgarian state loan of 30 million francs. Two groups competed: Länderbank, Banque de Paris et des Pays-Bas, Banque Internationale de Paris; and the Franco-German group Crédit Lyonnais, Breslauer-Disconto-Bank. The first group signed the contract on August 11, 1896. The French group imposed conditions for admission on the Paris bourse. The conditions effectively demanded that Creusot, instead of Krupp, receive the order for 120 guns. To appease the French, Bulgaria agreed to place a smaller order with Schneider on February 22, 1897. The 30-million-franc loan was subsequently issued through Banque de Internationale de Paris, Banque Paris, and the Austrian bank in June 1897, and 3 million francs were made available for arms orders.

In this manner Schneider scored its first success in Bulgaria as the government placed an order with Creusot for twenty-four howitzers (12cm) valued at 1.25 million francs. Creusot delivered twenty-three guns and carriages to Rustchuk in April 1901. Krupp had signed a contract with Bulgaria for an even larger order: 90 guns (8.7cm) and 23,000 rounds of shrapnel for the price of 2.135 million francs payable in treasury bonds.[57] Krupp still held the advantage, because its pattern already served as the standard one in use by the Bulgarian army. Nevertheless, Creusot's order served notice that the French had arrived as a serious challenger to the Germans.

The entry of the French in the Bulgarian arms market seemed to offer the possibility of dislodging Krupp. In an attempt to take advantage of the opening, and knowing that Bulgaria required financial assistance to make its arms purchases, the French tried to apply leverage through their financial houses in 1898, in effect attempting a kind of banking imperialism. With the Bulgarian government in need of loans, French bankers offered their services under three conditions. First, they did not want the Bulgarian government to grant any further concessions for railway construction. Second, the government would have to refrain from issuing any more treasury bonds. Finally, Bulgaria should give French firms preferential treatment in all industrial contracts, including for armaments. Although they were willing to accede to the first two French demands, the third proved too contentious for the Bulgarians, and the negotiations collapsed. In the end, the Germans provided 5 million francs repayable in six months on the security of the unissued bonds from the 1892 loan. The prompt German reaction probably owed something to Krupp's fear that if no action were to be taken, Bulgaria would be "driven by necessity to accept the French conditions."[58] It seems likely that Krupp was instrumental in obtaining this advance to prevent Bulgaria from placing orders with French suppliers.

For its infantry rifle Bulgaria continued to buy from Austria. In 1897 the Bulgarian government concluded a contract in Austria for 17,000 Mannlicher repeater rifles and 3,000 carbines. This purchase brought the total Mannlicher infantry rifles in the Bulgarian army to 160,000. Additionally, a representative of Maxim-Nordenfeldt arrived at Sofia to carry out trials of a 37mm quick-fire gun, hoping to induce the Bulgarian government to place an order. The efforts were rewarded when Maxim-Nordenfeldt received an order worth about 1 million francs.[59]

The 1890s also witnessed some modest activity for the Bulgarian navy. The well-kept naval arsenal at Rustchuk, a holdover from the days of the Ottoman Danube flotilla, comprised a small steam hammer, turning lathes, punching and riveting machines, bronze and iron casting appliances, moulds, and a furnace. At the start of 1897 a French naval officer, Lieutenant Mareau, was drawing up a course of study on the French model for Bulgarian flotilla officers. The only order was a small training vessel. A representative from the British firm Fairfield Shipbuilding and Engineering Company, having heard of the prospect of Bulgaria's ordering a ship, had come to Sofia in January 1897 to submit a tender before negotiations had advanced too far with the French company. Although the French had not yet sent in the drawings, the influence of the French naval officer on the spot would present a significant hurdle to overcome. Ultimately, in 1897 Bulgaria gave an order to the French Compagnie des Chantiers de la Gironde for a war vessel. The contract price was 1.6 million francs for the ship and 280,000 for armament.[60]

After the Russo-Turkish war, Russia abandoned Serbia in favor of Bulgaria. St. Petersburg had effectively told the Serbs to look to Austria as their Great Power supporter. The Obrenović dynasty, first under King Milan (1882–1889) and then his son King Alexander (1889–1903), complied and followed a generally pro-Austrian course. However, Serbian arms purchases did not conform to that orientation in foreign policy, and Austria reaped only a tiny portion of Serbian war orders. The Serbs approached Russia only to be rebuffed.

In April 1879 the Serbian government decided to acquire 120,000 new rifles. The artillery committee, working through the War Ministry, wanted to conduct comparative trials and solicited test models from suppliers. The French government, through its rifle works at Châtellerault, put forth the rifle model 1874 at an estimated cost of 8 million francs. The Serbian government conducted gun trials during the winter of 1879–1880 and determined that the Mauser rifle was the best. On June 19, 1880, the Serbian government officially accepted the Mauser rifle and the War Ministry entered into negotiations over price and delivery. With the endorsement of the Serbian War Ministry, the Serbian parliament (skupshina) approved 6 million francs for the Mausers on February 14, 1881. Thus, Serbia purchased 120,000 of the model 1871 rifles, designated as Serbian model 1878/80, but also known as the Mauser-

Koka or the Mauser-Milanović after the Serbian officer who modified it. The company delivered the final rifles on March 13, 1884. The Serbian government invited Paul Mauser to Belgrade, and he returned to Oberndorf on August 2 with a contract for 5,000 repeater carbines and 3,000 repeater rifles for the artillery service. Mauser completed this order in October 1885.[61]

During the war with Bulgaria, Serbia would have had great difficulties carrying on without Austrian accommodation. Austrian suppliers sold 300 boxes of Koka-Mauser ammunition for Serbs, and the Austrian government permitted the delivery across its territory of 3 million cartridges ordered by the Serbs from the Mackenzie Company in London. In December 1885 the Serbs ordered 1 million cartridges for Peabody-Martini rifles from the London firm John Mor and Son. Additionally, Serbia acquired 8 million cartridges from Austria and 6 million from Belgium. Following the peace treaty with Bulgaria, Serbia purchased and received 100,000 new rifles, 34 million rounds of rifle ammunition, 254 new artillery pieces, and 127,000 rounds of shell in 1886.[62]

To pay for all this equipment the Serbs had contracted a loan of 12.5 million francs for military purposes to be paid off in forty-one years. By 1887, however, Serbia found itself in financial difficulties. The government was unable to pay the troops for three months, and the state had to pension off officials. Yet the Serbian government had recently ordered 20,000 repeater rifles of Mauser pattern from Oberndorf in Wurtemberg. To economize further, the government also decided to exclude powder imports and to manufacture its own gunpowder under the direction of the Ministry of War. To that end a gunpowder plant was founded at Topolo. Serbia also created a state monopoly on gunpowder.[63]

Similar to the Bulgarian trade, Russian arms exports to Serbia originally consisted of older, secondhand equipment. In 1890 Serbia purchased 60,000 Berdan rifles and ammunition from Russia to arm its first and second reserve units. Serbian credit difficulties inhibited its acquisition of heavy siege guns for the frontier fortifications. In 1891 the Russian government offered to furnish the required guns on the same terms as the Berdan rifle deal of 1890. The Serbs accepted the offer, and thirty old breach-loading bronze siege pieces and six howitzers were shipped by Gagarin's steamboat company. The Serbs appreciated obtaining the weapons on favorable terms, but the Russians' tendency to sell obsolete weapons led to some grumbling among Serb military officials.[64]

The Serbs' desire for more up-to-date arms led to a series of negotiations with the Russian government. In November 1891 the Serbian prime minister Nikola Pašić expressed to the Russian military attaché in Belgrade, Colonel Baron O. Taube, Serbia's interest in acquiring the new Russian three-line Mosin rifle model 1891. Such an acquisition raised difficulties for Russia because the rifles were actually being manufactured in France for the Russian government. In pursuing rearmament, the Serbian government lacked domestic means to produce smokeless powder, cartridges, or rifles. The Serbs also faced financial restraints, and they knew that ultimately they would require a foreign loan for 25 million francs to pay for the rifles. The cost factor also meant that the weapons would have to be purchased in France since the French factories at Châtellerault already had the production line and thus could offer a cheaper price. In November 1892 the French signed an agreement with the Serbs for 80,000–90,000 rifles of the Russian model to be financed by a loan at 5 percent interest. The contract specified that the Serb rifles would be made in the same factory that was completing the Russian order. The Serbian War Ministry wanted 120,000 magazine rifles to keep up with Bulgaria.[65]

At this stage the production schedules between Russia and Serbia came into conflict. Because Châtellerault would not fill the Russian government's order for Mosin rifles before the end of 1894, the plant would not even initiate production of the Serb order until 1895. Since the Serbian government was asking for only 80,000 rifles, Serb officials approached the Russian government about the possibility for allowing part of the Russian order to be redirected to Serbia. The French appeared willing to help Serbia by authorizing the factory to complete the rifles for Serbia in July 1893. However, the Russians raised objections to the proposal, as the tsarist government's first condition for granting Serbia permission to order the rifles was that the Russian order not be hindered or delayed.[66]

Russian officials in St. Petersburg took up the problem of supplying Serbia early in 1894. In discussing the matter in January, War Minister Petr Semenovich Vannovskii, Finance Minister Sergei Witte, General N. N. Obruchev, and Graf Kapinst of the Asiatic Department voiced reservations about promoting Serb acquisitions. First, they noted that Serbia already owed Russia more than 2 million rubles, and this amount would increase to 6 million. Given Serbia's financial condition they deemed it unlikely that Serbia could meet the obligation. Although the Serb government had applied for a loan for rearmament, the recent moves against it on European stock market meant that it was not possible. In payment to the Russian government the Serbian government offered a monopoly on the sale of kerosene or spirits in Serbia. Political concerns caused reluctance more than financial ones did. The political struggle in Serbia could bring to power a pro-Austrian party that would then be well-armed contrary to Russian aims. Recognizing the sensitive nature of the situation, the Russian officials gave Pašić a technical explanation, namely that the need for Russia to complete its order at Châtellerault first made delivery to Serbia logistically unfeasible. The Russians also demanded better financial terms from the Serbs as a way to draw out the negotiations. French and German markets were hesitant to engage in Serbian affairs because of Serbia's poor financial situation. For example, in 1894 the Serbs prepared to order 150,000–200,000 Lebel rifles from France, but the skupshina rejected the French loan accord as too severe.[67]

Evidently Russia's delaying tactics worked, because by 1897 the Serbs were still trying to obtain the Russian rifles. In November 1896 Witte and Vannovskii began discussions about rifles for Serbia. By then the Russian production run had been completed, so the Russians looked on the Serb acquisition more favorably. On the financial side, Serbia's position had not improved much, since it still owed 1.7 million rubles for previous loans. From a political point of view, though, selling the rifles to Serbia could draw King Alexander closer to Russia. Also, Russia could benefit militarily since the sale could attract the Serb army to Russian plans in the Balkans, adding a possible 100,000 active soldiers. Based on these considerations, Russia decided to offer a new loan to Serbia through private Russian banks and to begin with a small order of 16,000–20,000 rifles. Then, depending on the course of events, the number of rifles could be increased to 100,000 over four to five years. By August 1897 the Russian government had worked out a tentative timetable for delivering 112,500 guns to Serbia from Châtellerault, beginning in March 1898 and finishing in July 1899.[68]

Having already waited on the Russians for six years, Serb patience was exhausted. In July 1897 the Serb government approached other potential suppliers. They sought financial backing in England and put out feelers to the French for an order of 100,000 Lebel rifles. The Russians, sensing that perhaps they had procrastinated too long, moved to regain

Serb favor by offering 30 million Berdan cartridges gratis. In the end, the Serbs turned elsewhere, and ultimately German Mausers became the rifle of choice. In June 1899 the government signed a contract with Deutsches Waffen-und-Munitions Fabriken (DWMF) for 90,000 repeater rifles with 600 cartridges each. The Loewe works in Berlin supplied the rifles while the 45 million cartridges came from a Belgian supplier. The contract included a provision for 40,000 additional rifles and plans to establish a Serbian factory to produce smokeless powder. To pay for the contract, the Serbian government arranged a loan of 3.2 million francs through the Wiener Union-Bank.[69]

This Mauser order proved problematic on both financial and technical grounds. The order cost 13 million francs. After remitting 2.5 million francs, the Serbs pulled 3 million francs each from the Military Invalid Fund and the Prisons Fund for the rifles. The government covered the remaining balance through loans from the Union Bank in Vienna, which arranged a promise of 3 million francs from Germany, and from the National Bank of Serbia. By February 1900, Loewe was delivering 6,000 rifles a month to Serbia. By November 1900 Loewe had handed over 56,000 rifles, but problems occurred. The Serbian Military Commission began complaining about the inferior quality. The Serbs found many of the rifles to be defective, and the commission refused to accept 64 percent of them because those rifles had already been rejected by the Chilean government. The cartridges supplied by the Belgian firm did not suit the rifle. The Loewe firm compounded the problem by rubbing down the markings and trying to present the defective rifles a second time. The Berlin firm also caused difficulties for the Serbs by dumping large numbers of weapons for inspection at the last minute, thereby making thorough examination next to impossible. Because of these practices, the Serbian government decided not to order from Loewe in the future, and in 1901 Serbia ordered 100,000 Mausers from Oberndorf.[70]

For its artillery the Serbian government sent a military commission to visit gun foundries in Paris and Essen to determine what artillery system to adopt in 1884. Two days after the commission's departure from Serbia, Major Hamilton Geary of the Royal Army, an agent of Armstrong's, arrived in Belgrade. Geary managed to have an audience with the king, who described the present Serbian artillery as barely worth the bronze from which it was cast. For that reason the Serbian government desired to reequip its artillery service with new field pieces. According to the king, they had no preconceived thoughts as to which system they should adopt, or to whom they should apply. They desired only to get the best gun and to obtain it at the most reasonable price. After meeting with Geary, the king instructed the commission to proceed from Paris to London and then to Newcastle, to include the Armstrong firm. In June 1885 the Serbian government granted the contract to the Cail factory in France under the direction of Colonel Charles Ragon de Bange. The order consisted of forty-five batteries of field guns (8cm) and seven batteries of mountain guns of the same caliber. The contract price for the batteries amounted to 6.5 million francs (£260,000). By December the first two batteries of de Bange guns had left France.[71]

In 1891, Handels Gesellschaft Bank of Berlin advanced Serbia a loan in the amount of 4 million francs. The Serbs responded by sending officers to Essen to arrange the purchase of heavy guns from Krupp for forts on the Bulgarian border. The Serbs also expressed interest in Gruson guns from Magdeburg and conducted firing tests. The Gruson trials were considered satisfactory, but no sale followed, due to insufficient funds. Unlike Krupp, Gruson was not willing to supply guns on credit.[72]

In February 1892 de Bange returned to Belgrade to negotiate with the Serbian government for nearly 200 heavy guns at an estimated cost of £1 million. If the skupshina gave its authorization the money would be provided through a foreign loan. Commenting on Serbia's difficult financial situation, British ambassador F. R. St. John reported, "The circumstance that Colonel de Bange, whose field pieces are exceedingly unpopular in the Serbian army, should be concerned in a question of further supplies has caused much surprise here and appears to indicate that the choice of firms competing for the privilege of arming Serbia is somewhat limited."[73] Finding no sale since the Serbs could not pay, de Bange moved on to Constantinople. Soon thereafter the parliament granted approval for a loan of 10 million francs (£400,000), sanctioned for army purposes. Once again Handels Gesellschaft of Berlin provided the financial support, this time for 8 million francs.[74]

By April 1892 the Serb government was contemplating another proposed loan of 40 million francs (£1.6 million) for artillery and supplies. The French embassy tried to secure an order for the heavy ordnance required to arm the newly constructed frontier forts. Still, the Serbs remained cool toward the French firm "as experience in the Serbian Army had not proved favorable to the de Bange guns with which their field artillery is at present furnished while recent avowals in the French Chambers as to the superiority of the Armstrong gun has here drawn general attention in that direction." The Serbs requested that F. R. St. John bring their interest to the attention of Armstrong to make a proposal. However, St. John evinced skepticism, reporting, "It remains to be seen, however, whether any financial group could be found to take up such a loan without securing to itself a concession for the whole supply of arms which it is intended to provide."[75]

St. John's assessment of French decline and British opportunity proved only partially correct. Although de Bange no longer commanded any support, another French firm fared much better. The Schneider-Creusot factory secured a modest order for 2.8 million francs in June 1897. Serbia contracted with Creusot for twenty howitzers (12cm), two howitzers (15cm), six mortars, and sixteen siege guns (12cm), all for fortress defense. The Schneider guns arrived in 1901 and formed the basis for a Serb siege gun battery.[76] Late in 1900 an agent from the British firm of Vickers-Maxim Nordenfeldt began talks with the Serbian War Ministry about supplying field artillery. King Alexander wanted to increase the complement of Serbian field artillery from forty-five to seventy-five batteries. A Serb commission had visited England and reported favorably on Vickers-Maxim. As potential collateral the government could offer 78,000 unissued bonds from the Serbian Treasury with a face value of 29 million francs (£1.56 million). Vickers could accept the bonds and place them on the financial markets in London and Paris. Vickers offered to supply thirty batteries at 15 million francs (£600,000) and would accept the bonds as payment.[77] Ultimately, the deal fell through, allegedly because of bribery. Austrian military intelligence had reported that the Serbian artillery commander charged with visiting Skoda, Krupp, Havre, and Creusot to assess quick-fire artillery had been bribed by Krupp.[78]

British firms faced greater challenges than their continental rivals due to geography and transportation. As Sir George Francis Bonham lamented to Lord Henry Charles Lansdowne in April 1902, "The difficulty with which English firms have to contend is that of transport. An instance has recently occurred in which some shrapnel shells sent as a sample by the firm of Sir W. Armstrong and Co. have not been delivered. The cases of shells have arrived having been shipped to Bourgas on the Black Sea and forwarded thence through Bulgaria; but for the charges and fuses it has been found necessary to ask permission of the Austro-Hungarian government which has not yet been granted."[79]

Serbian military orders in 1900 amounted to 9.67 million Austrian kronen. In the context of total Serb imports for the year, war orders amounted to 20.8 percent. From that figure, Germany accounted for 7.57 million kronen, of which cartridge orders claimed 3.58 million kronen and rifle purchases amounted to 2.27 million kronen. French military supplies comprised the remaining 2.10 million kronen, with Schneider's artillery costing 1.59 million. These purchases had a dramatic effect on the proportions of German and French trade in Serbia. Without the war orders Serbian purchases from France made up only 2.34 percent of total Serb imports, whereas with the inclusion of the war orders the French share increased to 6.56 percent. Similarly, the German portion increased from 16.6 percent to 27.85 percent with the addition of the arms trade.[80]

Greece's national poverty directly limited the weapons acquisition needed for national expansion. In the period 1879–1890 Greece considered a major expansion in naval power but settled on a more modest program. French and British shipbuilders sold the bulk of the vessels to Greece. Under the alternating governments of Prime Minister Trikoupes (1882–1885, 1886–1890) and Deliyannis (1885, 1890, 1895), Greece borrowed 630 million francs from 1879 to 1890, doubled its taxes, and entered bankruptcy in 1893.[81]

The French benefited first in 1880 when the Greek government approved a contract with Forges et Chantiers de la Mediterranée for two ironclads and two torpedo boats, to be paid in installments. The Greek government had arranged a temporary loan of 15 million francs from French sources on the condition that the money remained in France to pay for the military stores and equipment that the government had ordered there. The Greek government also had 17 million francs from an internal loan, so 30 million francs were available for military purposes. The Greeks favored Forges et Chantiers de la Mediterranée with another contract in December 1880 for construction of seven torpedo steam launches. At that time Greece also ordered four ironclad gunboats and a large number of launches from England. Greek officials planned to use the torpedo boats to harass Turkey in case of war.[82]

Beginning in 1883 British naval firms gained a larger share of Greek

orders even as the British government tried to discourage Greek purchases. Trikoupes wanted to order three or four large ironclads of 5,000
tons each, while Admiral Imbazi believed a smaller fleet of well-tried and
approved vessels would be more useful to the country. Another objection to such a class of vessels would be the absence of proper docks and
dockyards, which meant the vessels would have to be sent abroad every
time they needed repairs. In conversation with the prime minister, Sir
Clare Ford, the British representative, expressed "regret of Her Majesty's
Government that the idea should be entertained in Greece of expending
a larger sum of money on naval armaments than the financial position of
the country appeared to justify. I observed that His Excellency regarded
the question far more from a political than from a pecuniary point of
view, and he stated that even if Greece were to expend a very heavy burden on her resources, he considered that the acquisition of 3 or 4 powerful ships of war was a question of paramount importance to the proper
protection of the country."[83]

Considering it an impossible task to guard the extensive coastline efficiently with torpedoes, the prime minister preferred bigger, more formidable ships. He believed that if Greece had possessed a few powerful
ironclads back in 1878, then the country might have gained more from
the Berlin Congress agreement. Trikoupes noted that steps were being
taken for a floating dock and special attention to arsenals. Ford "reminded His Excellency that Turkey was an unhappy example how that
country had severely compromised her resources by the large outlay she
had made in the purchase of costly vessels which no one could attend
had been of much use to her, and I warned His Excellency to think twice
before laying out vast sums of money on ironclads which might prove in
the end little better than expensive toys particularly in a country where
a high standard of naval education amongst the officers had not yet been
fully developed."[84]

Ford's arguments did not change the prime minister's views, but the
king eventually decided in favor of more economical measures. In 1884
the Greeks were still contemplating buying some ironclads in England.
The cost, £1.5 million, was to be met by a loan. In the end, the Greeks
settled for six Yarrow torpedo boats (first-class), another English boat,
and an additional six first-class boats built in France. The Greek government abandoned the idea of constructing large ironclads and confined itself to increasing the fleet of torpedo boats. The king expressed his desire

to see Greece equipped with a fleet of torpedo boats and not hampered by costly ironclads.[85]

Throughout the 1880s the Greek government resorted to foreign loans to pay for procurement. The Greek budget for 1881 showed revenue of 51,481,560 drachmas, but expenditures claimed 113,852,722 drachmas, of which the Ministry of War required 61.5 million. Greece needed a loan to cover the deficit, and looked to Paris for a loan worth 80 million drachmas. In December 1885 a bill came before the Greek Chamber for a loan of 100 million drachmas (£3.76 million) to be applied only to army and navy expenditure for 1886. In 1887 the Greek government sought another loan, this one for 30 million francs to purchase ships. In November 1889 the Greeks ordered two cruisers from the French yard at St. Nazaire. The prime minister explained to the British that the cruisers were not a new idea, and his government would have purchased from British firms, but no British firm would accept bonds from the Greek government as payment. The St. Nazaire Company immediately had accepted this arrangement.[86]

In the summer of 1901 the Greeks pondered the possible purchase of a new cruiser from Armstrong. J. Falkner, Armstrong's agent, informed E. Egerton that Armstrong was anxious to sell artillery and ships to the Greek government, and the ambassador set the stage for an introduction to the king. Both the king and the prime minister wanted to purchase from Armstrong a fast cruiser that had originally been constructed for the Italian government. Egerton wrote to Salisbury:

> Nobody therefore can say that there has been any English pressure urging the Greek government to take what appears naturally to Mr. Corbett a financial risk. I give no opinion as to the wisdom of the purchase. I have warned Armstrong's agent of the possible future money difficulty. The Greek government do [sic] not pretend to pay more than eighty thousand pounds now of the price, which is between three hundred thousand pounds and four hundred thousands: the rest is paid at intervals with an interest charge. Curiously enough Armstrong's agent from inquiries at the National Bank did not appear apprehensive as to subsequent payments.[87]

Egerton speculated that Armstrong's motive was to tempt the king to order another ship from the English firm. Salisbury's margin note revealed the British government's disapproval of the sale. Regarding the order of a cruiser by the Greek government he wrote, "This is very ill behaved of the Greek Government and not quite right of Armstrong and Co. who have their hands full of orders from us, I believe."[88]

For their rifle and artillery suppliers, the Greeks had switched preferences from British and French suppliers to German and Austrian firms. The French Minister of War had promised to sell the Greeks 30,000 rifles in 1880, but difficulties arose with the looming threat of a Greco-Turkish war. The Greeks then turned to private French suppliers to arrange the purchase of 50,000 Chassepot rifles and ammunition. Even though the Greek government had paid for the weapons, the French government imposed an embargo on the shipment as the rifles made their way to Havre. The French embargo angered the Greeks. Finding their artillery and small arms materials lacking even before the military preparations of 1880–1881, the Greek government entered into extensive contracts with Krupp of Essen and the small arms factory at Steyr in Austria. Over the course of 1885–1886 the Greeks bought 40,000 rifles in Austria. By 1893 the Greeks possessed artillery pieces of Armstrong, Krupp, and Canet models. At one time, the Greek navy's armament consisted exclusively of Armstrong muzzleloaders. The old Armstrong guns went to coast defense while modern Krupp guns went on board the fleet. In 1896 the Greek navy selected Canet quick-fire guns to arm the *Psara,* and the French firm Forges et Chantiers de la Mediteranée received 1 million drachmas (£22,000) for the job.[89]

Romania worried most about Russia in this period. For protection, King Charles allied with Austria and Germany, and his government warmly accepted the German embrace of Krupp. Yet French financial participation also claimed a share of Romania's arms purchases.

Fearing that Romania could be caught in a conflict between Russia and Austria, and that in such circumstances it would be unable to obtain supplies and ammunition from abroad, the Romanian government began building an armaments stockpile. In 1882 the Romanian Chamber voted to purchase 20,000 rifles of the Steyr system in Austria and to spend 3.5 million francs for guns and other arms. Provisions also included some orders to be made at Armstrong. At the same time, the Chamber voted £1.1 million for the War Ministry, from which the military stores were paid. Because it was difficult to make contracts in advance, the Romanian government applied to the Chamber for special authority to enter into engagements for 1883–1884 within limits. Besides the £1,061,500 assigned to the War Ministry in the budget for the coming year, £160,000 was also requested. Romania ordered several batteries of guns, partly from Krupp and partly from Austria. It bought 70,000 more Steyr rifles and anticipated a further order.[90]

Surprisingly, British artillery had scored a victory in the Balkans. The unusual Armstrong purchases likely had resulted from bribery. In 1883 Armstrong furnished the Romanian War Office with eight batteries of jointed mountain guns complete with gun wagons and ten six-inch siege guns with hydraulic carriages.[91] In 1888 General Angelesco, the former minister of war, alledgedly secured considerable personal profits out of contracts within his department. John Broadwell, an American citizen and Armstrong agent, made revelations regarding improprieties, and as a consequence two Maican brothers (one a Romanian general and commander of the Romanian fleet, the other a colonel and member of the commission to which contracts were referred) faced arrest and trial.[92] On April 12, 1888, General Maican was found guilty of having demanded gifts from Broadwell for concessions granted to him for army contracts. He was sentenced to ten months imprisonment, was stripped of his grade, and was fined 400 francs for legal expenses. At trial Broadwell declared that he was offered a large sum of money if he would leave the country. "The sum offered was 8,000 pounds and Mr. Broadwell intimated his readiness to take 10,000 if he could be assured of obtaining a concession for providing powder for the Romanian army, in which case he would give 30,000 francs to Stephen Lakeman." Lakeman did not deny the accuracy of Broadwell's statement. Lakeman, an Englishman who had served in the Turkish army, was also known as Mazar Pasha.[93]

With Armstrong tainted by public scandal, Krupp's position as the dominant supplier for Romania remained secure. In October 1885 the Romanian army was conducting artillery trials in Bucharest against turrets. The Krupp and de Bange guns were employed against a turret manufactured by the German firm Gruson and another by the French firm St. Chamond.[94] Confident that his firm would hold the gun orders without much difficulty, Alfred Krupp instructed Friedrich Alfred Krupp to maneuver for the turret orders as well. Alfred Krupp wrote in 1887:

> Present my dutiful respects to the King of Romania . . . At the moment there will be no word of business. It will be very useful, however, if you prepare yourself for it,—not because of the guns, for they will certainly take those from us, and the King will be confirmed in his views in Berlin; but the sooner you have an opportunity of talking to him about the successes of our armor plate and of furnishing evidence of them,

the sooner the opportunity should be grasped, in order to get the people over there away from their ideas about Chamond and Gruson.[95]

King Charles played a critical role in shaping Romanian arms preferences. While the Extraordinary War Credit worth 12 million francs (£480,000) was being approved in 1896, discussions in the Romanian Chamber arose about the Austrian Mannlicher rifle. General Budishteans stigmatized the rifle as a "stick" that required improvements. The general's negative comments caused much consternation in the army, because what the general derisively referred to as "un baton" had the king's explicit approval. There had also been problems getting the right powder because the Romanian model had a smaller bore than the usual Mannlicher. With the king's backing, as of February 1901 Romania had 144,000 new Mannlicher rifles (109,000 purchased in 1896, 20,000 in 1897 and 15,000 in 1898). In May 1899 Romania was considering an order of 80 new batteries (480 guns); since 1898 officials had been conducting tests with Chamond, Krupp, and Hotchkiss pieces. The Romanian officer in charge, Colonel Perdikari, proposed to order from Chamond. However, the Artillery Commission endorsed Krupp, and there was speculation that Krupp's agent in Bucharest had been influencing the commission. Nevertheless, the king personally told Krupp's agent that, due to lack of money, the order would be cancelled. The government next turned to war surplus as a means to raise funds for rearmament. In 1900 the Romanian government received £140,000 for selling disused Martini-Henry rifles and old ammunition had been invested as partial payment for an order for thirty-four howitzers placed with Krupp and for cartridges placed in Austria for 6mm Mannlicher rifles. Because of the lack of money, King Charles had financed the artillery order out of his personal resources.[96]

Even though Krupp enjoyed the personal support of the king, the firm did not shrink from intimidating Romania if its business interests appeared threatened. Concerned about Bulgarian armament, Romania wanted its field guns to be compatible with Austrian ones. In an attempt to gain an edge, the War Ministry of Austria-Hungary had secretly sent Romania eighty boxes containing a new quick-fire gun (caliber 8cm) with carriage, grenades, and shrapnel. The first five boxes containing the gun parts were dispatched on October 2, 1901. In December 1901 Romanian authorities had tested the Austrian weapon and found it satisfactory enough to place an order. After hearing about this, Krupp threatened to demand immediate repayment of all the Romanian government owed the company if the Romanian War Ministry transferred its order. Krupp got his way, and in 1902 the German firm received contracts for artillery with ammunition amounting to £2 million.[97]

Whereas Krupp employed financial threats to preserve the firm's position, the French began to use their financial assets to break into the Romanian market. To pay for the Krupp order, Romania looked to German and French financial houses. Berlin bankers took 78 percent and French bankers took 22 percent in cartel of all loans and advances made to Romania. When Berlin opened a line of credit of £700,000 for Romania the French syndicate refused to take on its share, unless certain conditions were met. The French bankers demanded that the Romanian government spend £130,000 for guns in France. They also wanted to deal directly with Romania. The German house Disconto Bleichröder dominated Romania, and although it often collaborated with French banks, it and other German firms considered Romania their turf, and they often dictated conditions. The French now objected to passing through the intermediary of Bleichröder. Disconto Bleichröder could not accept French banks negotiating directly with Bucharest, as then German money would be used to pay French industry. Now the French government asserted that French banks should have an equal standing in Romania.[98]

In a private audience with the British representative J. G. Kennedy, the king of Romania expressed his indignation at the French banking syndicate's threat to refuse to cooperate with the Germans. The French refusal was significant because French investors held about £1.5 million worth of Romanian treasury bonds. In the king's view, the French government was to blame. "His Majesty said he intended to resist any pressure of this nature," and he hinted that he hoped an English syndicate would step in.[99]

Despite objections from the French government, French banks opted to continue to participate in lending money to Romania. At last, in March 1902, Romania received a loan for 250 million francs, part of which was designated for artillery rearmament with quick-fire guns. The majority of the loan (175 million francs) was earmarked to meet past obligations, since 1899, over a five-year period. The funds available for new artillery amounted to 40 million francs. Out of the 40 million, Krupp received 4 million francs as payments in 1902. Another loan followed in 1903. This one amounted to 185 million francs, of which the

French Comptoir National d'Escompte participated for 80 million despite opposition from Quai d'Orsay. As the first priority use for the funds, Romania paid 23 million francs for 450 Krupp guns.[100]

From 1878 to 1903 in the Balkans, the Austrians and Germans achieved hegemony in sales of rifles and artillery respectively. Krupp ruled the artillery markets in Bulgaria, Romania, and Greece. Only Serbia remained outside Essen's orbit. The Serbs, contrary to their pro-Austrian diplomatic leanings, bought French artillery and German rifles. The artillery purchases of these states closely mirrored those by the Ottomans, even though nothing comparable to the German Military Mission in Turkey operated in the other countries. The French mostly contented themselves with the financial side of the arms business in partnership with the Germans. Between 1898 and 1906 joint French-German loans provided 1.23 billion francs to the Balkan states, with Romania absorbing the lion's share with 1.064 billion francs.[101]

Each Balkan state endured financial sacrifices and assumed large foreign debts to shoulder the all-important defense burden. They all imported the arms and the money to pay for them. The region witnessed German and Austrian ascendancy in the arms trade, but the states arrived at those purchases by different paths. Romania, under King Charles, pursued an alliance with Austria and chose Austrian rifles so the two armies would be allies in a war against Russia. King Charles also personally opted for Krupp artillery and made sure he got it. The financial side often determined Bulgarian purchases as Austrian, German, and finally French loans imposed certain conditions for weapons contracts. Greek and Serbian choices had little to do with foreign policy and more to do with their poor credit ratings. French businesses kept their hand in through state loans to buy the armaments throughout the region, and by selling artillery to Serbia and Greece.

The Ottoman Empire bought less than it had in the previous era, and what it did purchase came almost exclusively from the Germans. By purchasing German hardware, Sultan Abdulhamid II hoped to wrap the aura of German military prestige around his armed forces and arouse German interest in supporting Turkey. Diplomatically, the Ottomans needed favorable terms with at least one European Great Power. Russia remained the chief adversary, and following the Congress of Berlin (1878), Britain and Austria-Hungary appeared unacceptable as partners since each had taken administrative control over parts of Ottoman territory. Only France, Germany, and Italy remained as potential protecting powers. Given the choice among these three states, the sultan selected Germany for practical reasons: The Germans had no border or potential area of interest with Ottoman lands, and German military prowess provided something worthy of emulation.

In Russia the tsarist naval and military bureaucracies, not the tsar, played the leading role in determining procurement. The War and Naval ministries oversaw committees that, in turn, had authority over all contracts and procurement. Ultimately, the Main Artillery Administration and the Technical Committee of the Navy handled the laborious procedures for setting technical specifications, negotiating with manufacturers, and determining delivery terms. Even though these bureaucratic entities lacked any legislative oversight, on the whole they functioned professionally, albeit ponderously, in evaluating the relative merits of weapons and warships based on cost and performance.

5

A Tale of Two Arms Races

Not all arms races are the same. The arms races in South America and East Asia occurred at roughly the same time (late nineteenth century) and largely involved the same private suppliers, but had fundamentally different dynamics. For Chile and Argentina the arms competition revolved around prestige. Each sought to claim itself as the most powerful republic in South America. Their arms race lacked an underlying strategic policy core. The race assumed the form of a naval race because warships represented the highest-profile items a state could acquire, and therefore the posturing could be measured in the relative tonnage of the biggest and most modern naval assets acquired. Lacking a serious strategic underpinning, the Chilean-Argentine naval race could be resolved peacefully and quickly since no vital issue was at stake for either side.

Across the Pacific, Japan pursued a strategic arms buildup of a Clausewitzian sort. Japan initiated wars first against China and then Russia as an extension of its policy goals to dominate Korea and establish itself on the Asian mainland. Japanese leaders developed a specific interest in the Korean peninsula as the key to Japanese defense and Japanese expansion into China. Japan followed concerted plans of industrial development and arms imports, especially naval, to obtain the necessary war tools to achieve its aims. Thus, the Japanese integrated their armaments purchases into their overall offensive imperialist strategy, making themselves more formidable and much more likely to start a war.

The differences between the two arms races were also reflected in the methods of financing. The South American states relied on foreign loans to purchase their armaments, and this made it possible for foreign creditors to apply pressure to facilitate the peaceful end of the arms race. Japan paid for its arms buildup by mobilizing domestic resources through higher domestic taxes. The increasing taxes and budgetary allocations for the army and navy received strong legislative and popular support because they were perceived to be in the national interest of Japan. In this way the national will buttressed the Japanese arms race, and it was less subject to manipulation by outside business interests.

As was the case in Eastern Europe and Ethiopia, regional conflicts stoked the arms business in South America. However, unlike those other areas, in South America the naval trade figured more prominently. On the Atlantic side the spark for the arms business came from within the continent. The rivalry between Brazil and Argentina for preeminence in South America served as the central tension. On the Pacific side Spanish naval intervention provided external stimulation for Peru and Chile to acquire newer, more advanced naval forces in competition with each other. For Argentina, Brazil, and Chile direct participation in wars provided the initial desire for arms purchases, but emerging from those wars with newfound strength opened up possibilities for grander ambitions than defense in time of war.

On the Atlantic side, the Argentine-Brazilian rivalry over the River Plate basin overshadowed relations for most of the nineteenth century. The War of the Triple Alliance (1864–1870) marked the opening of a new era in the development of the arms business, especially the naval trade. In that war Argentina, Brazil, and Uruguay had their hands full for six years trying to subdue Paraguay. General Bartolome Mitre of Argentina had entered the war with the hope that victory would place Argentina into "the first rank among the South American republics," but by 1868 the poor showing of Argentina's forces compared to Brazil's led critics to accuse Mitre of turning Argentina into a "mere prefecture" of the Brazilian empire. Due to the lessons of that conflict, Argentina's leaders, first Mitre and then Domingo Faustino Sarmiento, determined that Argentina needed increased naval power to assure its position vis-à-vis Brazil.[1]

Brazil occupied pride of place in South America in the 1860s. Besides having the largest population and territory on the continent, Brazil possessed the largest navy in Latin America. Of the forty-five vessels in the Brazilian navy, thirty-five were steam-powered, including fourteen ironclads. Brazil built some of its own river monitors and gunboats, but also had bought British-made warships and British artillery in the 1850s. During the war with Paraguay, the Brazilian army bought and deployed state-of-the-art Krupp guns in 1867.[2]

The Argentine navy did not have true warships built until the 1870s when President Sarmiento (1866–1874) began developing the Argentine navy in earnest. In the summer of 1869 the Argentine government approached Laird Brothers at Birkenhead about designs and estimates for armor-clad turret vessels and unarmored gunboats. By December 1872 Argentina was prepared to order two ships of each class. The gunboats were priced at £21,500 each and were to be armed with Armstrong artillery. The battleships *La Plata* and *Los Andes* cost £85,600 apiece. In February 1874 the two gunboats for Argentina constructed at Birkenhead were ready to flag, and the battleships arrived in 1874 and 1875. These ships, known as "Sarmiento's Fleet," were intended for service along Plate tributaries in case of war with Brazil.[3]

Argentina's interest in developing its naval forces beyond a mere river and coastal protection force became evident in 1878 when the Argentine government made inquiries about construction of war vessels in England. For the first time Argentina wanted to buy a seagoing ironclad from the British government. The British government itself did not sell any ship, but Argentine authorities did enlist the skills of W. H. White, assistant constructor of the Royal Navy, to consult for them on his private time. Specifically, the Argentines wanted White's expertise respecting designs for an armored warship.[4]

Fears of an Argentine–Brazilian arms race were palpable in 1880–1881. Sir Horace Rumbold, British ambassador in Argentina, reported to Lord George Granville that "there has unquestionably been a scare with regard to Argentine armaments and Brazilian counter armaments," but in Rumbold's estimation Argentina's navy was not a very formidable force.[5] Rumbold noted that Argentina's purchase of the more than 4,000-ton ironclad *Almirante Brown* being constructed in Britain (Samuda) did not constitute a cause for alarm, in his assessment, although *Almirante Brown* was "by no means a small vessel." He continued:

> I think it likely that, by the time the last installments of her cost and the necessary expenses to navigate her to the River Plate have been paid a large amount of the money voted by the legislature within the last four months for the Naval and Military forces of this republic will have been absorbed . . . The ships themselves, I fancy, are not maintained in an ef-

ficient condition never being docked or otherwise examined . . . In very few years these vessels will probably fall into the same state of decay as the ironclad vessels of Brazil's appear to have now reached, judging by those seen at Rio de Janeiro.[6]

Rumbold's analysis of Argentina's limited financial resources proved accurate. Between 1874 and 1885 the government spent £565,200 for new warships abroad, of which the *Almirante Brown* cost almost half (£270,000).[7] The Argentines were modernizing their equipment and rearming their troops with better weapons—Remington rifles—but "their activity in Naval and Military matters lately displayed is great only by contrast with their previous inaction, especially on the part of the navy."[8]

The Brazilian government perceived Argentina's armaments modernization as an aggressive challenge. In response the Brazilian Chamber voted 5 million milreis to increase naval forces of the empire, although most of the money voted was needed to render their current ironclads serviceable. Then, in 1882 Brazilian naval minister Sabina Elay Pessoa proposed a Brazilian naval reform plan for new ships over four years. The proposal included adding two ironclads, two cruisers, two ironclad monitors, fourteen gunboats, and twenty torpedo launches, with the smaller vessels being acquired within eighteen months. At the time, the Brazilian fleet consisted of four ironclads, seven cruisers, eleven gunboats, two transports, one frigate, and five torpedo launches. However, three ironclads, two steam corvettes, three gunboats, and one transport had been condemned as useless. The usable fleet possessed 118 cannon and a total of 26,071 tons. The ironclad *Riachuelo* was being built in England while the second-class cruiser *Imperial Marimhezo* was under construction at Pont da Area using steel provided by Krupp. By 1884 Brazil had eight ironclads and six Thornycroft torpedo boats that ejected Whitehead torpedoes.[9]

By now Argentina's navy had overtaken Brazil's. In the mid-1870s Argentina had attained parity with Brazil in weapons and warships, but by the 1880s Argentina had pulled ahead. Simultaneously, a potentially more powerful rival was rising to the west. In the late 1870s Argentina began to develop an oceangoing navy to match the Chilean fleet. Chile already had two 3,500-ton frigates, but by 1885 Argentina possessed only a single vessel of 4,200 tons. The other three ironclads in the Argentine navy were smaller vessels of about 1,500 tons each.[10] Argentina feared an alliance of Chile and Brazil. If anything, the state-of-war readiness and the great victories achieved by Chile on the west coast made Argentina still more wary.

Spanish gunboat diplomacy on the Pacific coast of South America exposed the countries of the region to the potency of modern sea power. In 1863 a dispute between Spanish workers and Peruvians ended in violence. On the initiative of the naval commander on the spot, a Spanish naval squadron in the area demonstrated its displeasure with Peruvian legal proceedings against the Spaniards by taking control of Peru's guano-rich Chincha Islands and would relinquish them only after getting legal satisfaction. Since the Spanish squadron easily outmatched the combined fleets of all the Pacific coast republics, Peru gave in to Spanish demands in January 1865. Meanwhile, resentment against Spanish high-handedness boiled over in Chile, and when Chilean officials refused to sell coal to the Spanish squadron, the Spanish admiral presented an ultimatum to the Chileans that pushed them to declare war September 24, 1865. To teach the Chileans a lesson the Spanish squadron bombarded the undefended Chilean port of Valparaiso. Common concern about the Spanish coalesced into a four-power alliance of Peru, Bolivia, Ecuador, and Chile that finally drove off the Spanish. In the aftermath, Chile and Peru gained a new appreciation for the overriding importance of sea power in any conflict along the coast.[11]

Peru and Chile then made moves to enhance their naval power through imports. Peru moved first by acquiring two powerful new ironclads, *Huascar* and *Independencia,* from British yards in 1866. These two ships gave Peru complete domination of the Pacific coast because they proved impervious to all the wooden ships of the other regional navies. Not to be outdone, Chile placed orders for two more powerful ironclads. In 1866 Chile sent an agent to the United States to buy the ironclad *Vandenburg,* but the U.S. government rejected the offer. The Chilean government still held out hope that it could acquire a two-turret ironclad lying in a Brooklyn dockyard through a private U.S. citizen who had lived in Chile. However, by the end of the year the Chileans dropped their potential orders for ships and war supplies in the United States because of the expense. In 1869 Chile finally obtained two British-built armored corvettes (1,690 tons each), the *O'Higgins* and *Chacabuco,* which Santiago had ordered in anticipation of the coming war with Spain. As a general rule, Chile bought British-built ships, and all the Chilean warships in commission in 1874 carried Armstrong guns. In 1872 Chilean

President Federico Errázuriz Zañartu contracted with Earle's at Hull for two reconstructed 3,500-ton oceangoing ironclads, the *Blanco Encalada* and *Cochrane*. The delivery of these state-of-the-art armed frigates by 1875 brought the naval advantage to Chile, since no gun in the region could penetrate these armored vessels. Chilean leaders could now imagine projecting their naval power from the Straits of Magellan all the way to Panama.[12]

The outbreak of the War of the Pacific in 1879 sent Peru and Chile scrambling for additional naval units. In May 1879 the Peruvians were anxious for the purchase of an additional ironclad, which would place their navy in a position of superiority over that of Chile. The Peruvian government had agents in Europe who were about to apply to the Ottoman government for the sale of one or more ships of its fleet. The Chileans approached Britain diplomatically to enlist British aid to prevent a possible Peruvian purchase. Britain remained neutral, although British ambassador St. John expressed skepticism that the Turks would part with any of their warships except for an extraordinarily high compensation. While the Turkish negotiations were unresolved, Chile and Peru each sought to buy ironclads in France. In 1880 Peru was still looking to buy ironclads. A Peruvian agent trying to buy two ironclads from Turkey sought Sir Henry Layard's assistance and influence. Layard, the British ambassador to the Ottomans, refused, noting that since Peru and Chile were belligerents Turkey would be breaching neutrality. Also, Layard was certain the sultan was determined not to sell any of his ironclads. Again Britain refused to get involved, as to do so would violate neutrality in the conflict between Peru and Chile.[13]

Frustrated in their efforts to acquire armed ships through official channels, the Peruvians explored private channels. In 1881, two vessels, *Socrates* and *Diogenes,* had been built by Howaldt and Company by private contract at Kiel, supposedly for the Greek government. The Greeks had not ordered them, however, and suspicious that they may have been for Peru, German authorities detained the ships in port, since they were designed to carry guns.[14] Next, Peru tried to get two ships sold to Henry Lambert of London. Lambert said he was supposed to sell the ships in question to the French government, but the Chilean Legation discerned the hand of Peru and requested that the vessels be detained.[15]

On the other side, Chile, too, pursued official and unofficial means to acquire warships. Chile had placed orders with British yards, especially

Armstrong, Mitchell and Company. To protect British neutrality, British customs officials informed Mitchell that the firm was not to move any Chilean vessel from the company's slipway. This proved problematic for the company. As Elswick explained to British authorities, the vessel in question was one of six vessels, and five were intended for China. Of the three ships building at the moment, however, Armstrong could not determine which one specifically was destined for Chile. Armstrong did inform the Chilean government that its ship could not be delivered until hostilities had concluded.[16]

Unable to obtain delivery of its warships, the Chileans tried to sneak around the British restrictions. They devised a plan to sail their ship out of England without her guns on board on the grounds that without the guns it was no longer a warship. Then a Chilean steamer carrying rifles, Gatling guns, and ammunition entered the port of Newcastle. The steamer, *Almirante Castle,* proposed to take on cargo at port, including the Armstrong guns belonging to the unarmed warship. The Peruvian Legation in Britain caught wind of the scheme and reported it to the British government. As the head of the Peruvian Legation put it, "This attempt to avoid the interference of the English Government on the pretense that the vessel is not armed, can, I think hardly avail, since there is no doubt from the construction of the vessel, she can only be intended for no other purpose than that of warfare."[17]

Chile ordered the construction of a new protected cruiser from Armstrong in 1881, an act of great importance for the future naval race in South America. Because the war with Peru was still in progress, Chile could not hope to receive the ship until the end of hostilities. Nevertheless, the arrival of this newly ordered ship, *Esmeralda,* was meant to ensure that Chile would command naval mastery against its rivals. *Esmeralda,* designed by George Rendel of Armstrong, had been paid for by popular subscription in Chile. Two years before its arrival, the Chilean press was already extolling the merits of the powerful new vessel.[18]

Chile versus Argentina, 1888–1902

As author William Sater has observed, Chile had emerged from the War of the Pacific not simply as the dominant South American power on the Pacific, but also as a hemispheric naval power capable of posing a challenge to the United States. Indeed, thanks to the *Esmeralda,* Chile's navy was more powerful than the U.S. Navy. Chile's ironclads could easily handle every wooden vessel in the U.S. fleet. In assessing the *Esmeralda,* the *American Army and Navy Journal* stated that this one ship "could destroy our entire Navy, ship by ship, and never be touched."[19] Chile considered the United States a rival, but Santiago possessed the stronger navy, and Chile's leader was determined to preserve the advantage for his country. Following the war, Chile continued to update its fleet by acquiring powerful additional vessels. President José Manuel Balmaceda stated, "Chile should be able to resist on its own territory any possible coalition, and if it cannot succeed in attaining the naval power of the great powers, it should at least prove, on the base of a secure port and a fleet proportionate to its resources, that there is no possible profit in starting a war against the Republic of Chile."[20] As proof that Chile meant business, during the 1885 isthmian crisis Chile sent *Esmeralda* to Panama to restore order and stop the U.S. fleet from annexing the Colombian province.

In South America the stage was now set for an intense arms and naval race between the two rising powers, Chile and Argentina. Santiago moved first in the late 1880s when President Balmaceda implemented plans to expand Chile's armed forces with significant new purchases of artillery and warships. These acquisitions went beyond simple modernization of equipment, representing an aspiration to project power regionally beyond mere defense. Conflicting boundary claims added to the tension. Argentina maintained that the border should be from the highest summit of the Andes. This meant Argentina was looking for an entrance to the Pacific in the south. Chile feared for its commercial and strategic position if Argentina should become a bicoastal state.[21]

In late 1888 President Balmaceda set in motion plans to expand and modernize Chile's military forces. At the time the Chilean army numbered roughly 12,000 men compared to Argentina's 9,000 and Brazil's 13,775. The Chileans were armed with Gras breech-loading rifles made by Steyr, while their artillery was a mixture of Krupp, Armstrong, Nordenfelt, and Gatling guns. Those in the cavalry were armed with Winchester repeating carbines. Chile was now looking to procure 100,000 rifles, 100 field guns, 8,000 sabers, and 5,000 carbines. Beyond these quantities, by 1890 Chile proposed in time to have 75 artillery batteries, one for every 1,000 men. Such quantities would be enough to equip an army of 80,000.[22]

The beginning of Chile's exertions to become a major naval power can be seen in the budget figures for the 1880s. As a general rule, in any given year army spending almost always exceeded naval spending for most countries in the world. From 1885 to 1887 Chile's annual total military expenditure (combined military and naval) as a percentage of Chile's budget averaged 15.64 percent, while naval budgets alone accounted for 5.83 percent. The figures for the next three-year period (1888–1890) reveal dramatic naval increases. While the annual total military expenditure on average increased to 20.29 percent, the average annual naval budgets almost doubled, to 9.41 percent. Specifically, in the budgets for 1889 and 1890 Balmaceda's government expended almost as much for the navy as the army. Even more important, the naval percentages increased even as the total expenditures increased, so the navy claimed a bigger slice of an expanding pie.[23]

Over the course of 1889–1890 Chile expanded its navy through contracts with French firms. In France construction of the *Capitan Prat* was up to the armored deck. Although the ship was due to be completed in twenty-seven months from the contract signing on April 18, 1889, the builders expected to launch it by the end of 1890. The steel armor came from Creusot, and the ship cost £391,000. Two protected cruisers, *Presidente Errazuriz* and *Presidente Pinto,* were also being built in France. The contracts were dated March 20, 1889, and each ship cost £105,180 for the hull and £20,000 for guns.[24]

The predilections of President Balmaceda directly caused the decline of British naval sales and the sudden rise of French orders. J. G. Kennedy, the British representative in Santiago, feared that British commercial enterprise in Chile was being superseded. Of five warships being built for Chile, the three largest were French. English firms had received orders for only two torpedo boats. The former Chilean prime minister told Kennedy that Admiral Juan José Latorre went to Europe with the intention to order all warships in England, but he had received orders from the president to give the preference to French shipbuilders. "The above orders, said Señor Lasterria, who was Prime Minister at the time, were simply due to what the President considered to be the unfriendly and even offensive action toward Chile of Her Majesty's Government in regard to the claims of the Peruvian bondholders." Kennedy also took into account that Antuney, the Chilean representative to London and Paris, and Admiral Juan José Latorre both resided in Paris where they had family ties. Both men had also been "the objects of much attention from interested parties,"[25] meaning that bribery was not unlikely.

In January 1891 Chilean officials began complaining about the French. Vienna, the recent first secretary of the Chilean Legation in Paris, expressed to the British his view that the Chilean government was beginning to realize the "small value which could be placed upon the friendship for Chile professed by the French government with a view to obtaining concessions for warships, railway material etc., and that scarcely any contract which they had entered into had not been subsequently disputed in some form or another."[26] Vienna feared that to bring pressure to bear in connection with the Dreyfus claim—Auguste Dreyfus of Paris had signed a contract with Peru for a guano concession before the War of the Pacific, but Chile's annexation of the territory put that concession in jeopardy—the French government might possibly attach the three ships under construction in France at a cost of nearly £1 million, of which nearly half had already been paid. If the French government did take this high-handed measure against the Chilean government, Chile would receive no sympathy from England. Although nine British firms had tendered for construction of these ships, the British had learned from a Chilean informant that Chile had decided to give the contracts to the French Compagne des Forges et Chantiers (La Seyne). As an added insult, the Chilean agents in France had handed over designs prepared in England, which were made use of by the French constructors.[27]

The Germans were also trying to ingratiate themselves for Chilean naval orders. The German ambassador, Felix Baron von Gutschmid, strove to get a contract for the Vulcan Company. At Gutschmid's suggestion, Vulcan appointed Herr Schmid-Von, a leading German merchant of Valparaiso, as its agent in Chile. Kennedy reported that "Gutschmid displays an ostentatious benevolence for the President and has, I know, complained that Chile failed to show by deeds any real appreciation of the benevolent political attitude of Germany towards this country."[28]

The Germans had more success with artillery sales since the Chileans had previously manifested their preference for Krupp. In the bidding process for coastal artillery Krupp gained an order for ten coastal guns in 1889, even though Schneider and Armstrong had both tendered below Krupp's price of 3.2 million marks. Here financing played the key role since Krupp had arranged a loan for Chile from Deutsche Bank and Mendelsohn and Company for 30.6 million marks to cover the artillery contract and orders for the Chilean railway.[29] In 1890 an unequal field gun contest pitted the latest Krupp model against the French de Bange 1877 model produced by Cail et Compagnie near Paris. The de Bange cannon used ordinary powder, while Krupp employed smokeless powder. Adding to Krupp's advantage, the firm's agent, Albert Schinzinger, oversaw operations during the tests whereas the French firm did not have an expert to represent it. Not surprisingly, the Krupp outperformed de Bange and on March 17, 1890, the Chilean Commission voted for Krupp. To squelch any misgivings about the overwhelming superiority of the Krupp gun, the president of the Testing Commission dismissed the minority opinion against Krupp and presented the findings of the commission as unanimous. In April 1890 Chile signed a contract worth 1.6 million marks for 82 Krupp pieces and 25,000 shells.[30]

In 1891 a struggle between the president and Congress over constitutional authority erupted into civil war. The conflict had been brewing since December 1890, when the majority in Congress declared Balmaceda unfit and tried to remove him from office. In his presidential message to Congress, Balmaceda had asked for £100,000 to acquire certain ships for the Chilean navy. The presidential request was referred to a Senate committee that had placed itself in communication with a naval committee. Congress was unwilling to approve expenditure on the eve of its dissolution, and Congress wanted to complete fortifications at Valparaiso and enroll more naval cadets before expanding the navy. The legislative session of 1890 ended with Congress refusing to pass the appropriations bill for the coming year and specifically refusing the expenditures for the armed forces. At the start of January 1891, the commander of the Chilean navy, Captain Jorge Montt, preempted Balmaceda's plans to neutralize the navy. Montt sailed the fleet out of Valparaiso with prominent members of the congressional leadership on board. With the conflict out in the open, the navy backed the congressionalists while the army supported the president.[31]

Both sides recognized the significance of the arms trade in determining the outcome of the armed struggle. President Balmaceda's government struggled to take possession of the two cruisers being built in France. The congressionalists sent Auguste Matte and Augustin Ross to Europe as confidential agents to represent the provisional government at

Iquique in Chile. Their mission was to block the procurement of ships and weapons by President Balmaceda. Balmaceda's government had already contracted to purchase naval guns and shells from Armstrong, and those munitions were about to be exported. In fact, six large Armstrong guns were about to be loaded onto a Chilean cruiser for delivery to the president. Matte and Ross urged the British government to prevent the weapons shipment on the grounds of British neutrality in the conflict. A similar sentiment came from Colonel John T. North, who wrote to Salisbury in May 1891 urging neutrality since partisanship could threaten the large English interests existing in Chile.[32] In the end Britain did observe neutrality in the conflict.

France and Germany, too, remained neutral. Balmaceda's government experienced a major blow from the French. In June the Civil Tribunal of the Seine ordered the sequestration of the warships built in France for the Chilean government. Matte and Ross had applied to the court for delivery of the vessels in question. The French ruling meant that the ships would remain chained in the harbor until the end of hostilities. Behind the scenes, Matte and Ross got the ships *Presidente Errazuriz* and *Presidente Pinto* detained by offering to pay the builders the installments due, to pay them punctually, and by guaranteeing the 5 million francs to be paid monthly while leaving the vessels in possession of the French firm until the termination of the contracts. In Berlin, German bankers refused to furnish the final portion of a loan due to the Chilean government. Balmaceda and his supporters had counted on the two French cruisers to suppress the revolution within a few weeks. Ricardo Cruzat, the Chilean minister of Foreign Affairs, declared that "Chile would never again give orders to France for the construction of ships."[33]

The congressionalist victory in the Chilean civil war had a profound effect on Chile's choice of arms suppliers. On the eve of the civil war in 1891, the government had adopted the Austrian Mannlicher rifle and German Krupp artillery to refortify coast muzzleloaders with breechloaders and for field guns, and placed naval orders with France.[34] Over the next decade Krupp would retain its hold on Chilean artillery, and German preeminence in the realm of military supplies would be reinforced when Mausers made by the Loewe firm in Berlin replaced Austrian-made Mannlichers as the main Chilean rifle. The German influence in military reforms and equipment was part of the "Prussianization" of the Chilean army. As authors William Sater and Holger Herwig describe the results, the victories of Emil Körner, a Prussian artillery officer hired to train the Chilean army in 1885 who supported the congressionalists during the civil war, paved the way for a massive influx of German war orders. Less than a year after the end of the civil war the German house of Vorweck and Company became the official Chilean representative of Ludwig Loewe in Berlin. Through Vorweck, Loewe began to pressure Chile to buy Mauser rifles.[35] By bringing in Prussian instructors, Chile set the bar for military modernization in the region, and in no other Latin American country did the Germans have such direct command within the army as they did in Chile.[36]

Argentina's decision to update its equipment conformed to the general global pattern of military technological change to magazine rifles that had begun in the mid-1880s. The Argentine government decided on rifle modernization in 1890. By then the old Remington rifles were showing their age. Colonel Pablo Riccheri headed the Argentine Arms Purchasing Commission during the 1890s. In 1891 Riccheri led a commission to purchase arms in Europe, and he selected the Mauser model 1891 manufactured by Loewe. As a temporary measure the Argentine arsenal would modify the Remingtons to take Mauser 7.65 ammunition until the new Mauser rifles arrived. In 1892 Argentina placed its order for 100,000 Mausers and 20,000 carbines from Loewe. Peru also ordered Mausers at this time. By 1897 Argentine had ordered 150,000 rifles to supply its army and navy.[37]

Argentina's suspicions of Chile helped propel the decision to modernize with Mausers. Fearful that Chile's 1889 armaments expansion would strengthen Santiago's hand in the ongoing border dispute, Argentina did not want to be left behind militarily. As part of the contractual terms, Argentina insisted that Loewe could not sell its rifles to any other Latin American nation, especially Chile or Brazil, for five years.[38]

For the Germans, standard operating procedure involved capitalizing on political tensions to gain war orders. To generate more military orders for German firms, Schinzinger, Krupp's agent, purposefully leaked information about Argentina's Mauser order to Boonen Rivera, then Chile's military attaché in Berlin. For his part, Loewe granted Rivera a tour of his rifle plant where the Chilean officer could clearly view all the new rifles bearing the Argentine crest. Then Körner moved in to secure the order. In November 1894, in his capacity as vice president of the Chilean

Arms Purchasing Commission in Europe, Körner urged President Jorge Montt of Chile to order 50,000 rifles and 20,000 carbines from Loewe. However, the German maneuvers almost backfired. Because Loewe was contractually prohibited from selling Mausers to anyone but Argentina, Chile took bids for rifles from the Belgian Fabrique Nationale in Herstal and Steyr. The Germans had inadvertently generated business for their competitors.

Only Argentina's financial difficulties saved Loewe. When Argentina proved unable to pay on time, Loewe used the breach of contract to pursue the Chilean order. Argentina did scrape together enough money to purchase 35,000 rifles, but too late to preempt Loewe's sale to Chile. In response to competitive pressures and complaints within the Chilean military, Loewe had to drop the price and promise a more rapid delivery schedule. Loewe's price of 77 francs per rifle remained higher than Herstal's quote of 70 francs or Steyr's at 69 for its Mannlicher. To eliminate the Belgian threat, Loewe brought legal action against Herstal for patent infringement and ultimately bought control of the firm in 1896. To neutralize the Austrians, Loewe's firm promised a more modern rifle, model 1893, in the future. Finally, Chile signed Loewe's contract and bought 50,000 Mauser rifles and 10,000 carbines for 3.2 million marks.[39]

German military influence vanquished the Austrians. For its last sale to Chile Steyr could muster a contract to produce only 54 million Mauser rounds and 8 million Mannlicher rounds. In 1895 Loewe received additional contracts for 30,000 Mauser rifles and 10,000 carbines. In 1897 the DWMF, by this time the controlling owner of Herstal, gained a Chilean ammunition contract for 60 million Mauser bullets. After DWMF's new Mauser model 1898 had been accepted by the Prussian Rifle Testing Commission, Chile acquired that model in 1901 for 3.8 million marks. Once again the Chilean rifle order came about after the Germans had provided the Chilean representative in Berlin with evidence of an Argentine order.[40] Having just recently adopted the Mannlicher rifle in 1892, Chile's navy discarded it, and both the army and navy were using the Mauser by 1897. Chile itself also exported some weapons, selling 5,500 of its surplus Mannlichers to Ecuador.[41]

The Germans continued their strategy of manipulating the Chilean–Argentine rivalry. Schinzinger, having sold more Krupp guns to Chile in 1892, moved on to Buenos Aires and sold Argentina six batteries of

Krupp field guns. Having personally fostered arms sales in Argentina, Schinzinger then told the German ambassador, Gutschmid, to alert Chile that "Argentina's warlike preparations were directed against Chile."[42] Körner, too, repeatedly spread rumors of war with Argentina to gain new Chilean orders in 1895. In the years 1890–1898 Krupp sold Chile 322 guns for 6.5 million marks. Chile acquired still more Krupp artillery when in December 1901 it purchased seventy-three field guns for 2.1 million marks. This order demonstrated the height of German influence since the Krupp guns cost 25–50 percent more than the bid prices of Krupp's competitors (Canet, Skoda, Bofors). Moreover, the others offered superior recoil.[43]

For all the institutional advantages for German influence and lobbying in Chile, Argentina proved a bigger customer. Argentina, with 683 guns, bought twice as much as Chile's 330 guns in 1887–1899, while Brazil purchased 208 Krupp pieces. Krupp's contracts with Argentina in 1885–1895 amounted to 60 million marks, whereas those for 1896–1898 alone totaled 12 million marks for 351 guns. In Argentina Krupp orders also made gains against British interests in the realm of naval guns. Previously, Armstrong had supplied most of the armaments for the Argentine navy. In 1890, however, Argentina moved ever so slightly in favor of Krupp guns. Consequently, Chile replaced the two Armstrong breechloader guns for the *25 de Mayo* with Krupp breechloaders, and later the *Independencia* and *Libertad* each received two Krupp breechloaders.[44] These naval orders represented only a modest gain for Krupp.

In the 1890s British firms again soared to predominance in the Chilean naval market. The composition of the Chilean navy in 1897 reflected a British monopoly. Excluding the three French warships launched in 1890, all the new acquisitions came from British yards. Armstrong built the main naval units, such as the four cruisers. Laird provided the majority of smaller vessels, including torpedo boats and destroyers.[45] The root cause of British dominance lay in the professional preferences of the Chilean Admiralty. Chile's naval commanders reasoned that since Britain was the premier naval power in the world, British yards must produce the best warships. This attitude came through clearly in 1895 when Ramon Subercasseaux, Chile's minister plenipotentiary in Berlin and Rome, conveyed an Italian offer to build cruisers for Chile. Much to the minister's dismay, "The rejection came after the government consulted the navy. (Apparently) the navy did not consider a ship to be a ship unless it came from England."[46]

Argentina also purchased from British yards in the 1890s, but the Argentine navy did not exhibit such an exclusive favoritism toward British-built warships. For its river defense in 1890 Argentina received four first-class torpedo boats, two third-class torpedo boats, and two second-class boats from Yarrow; twelve more Yarrow boats were on the way. Beginning in 1895 Argentina made major purchases from Italian yards. From 1890 to 1898 Argentina spent £4,534,800 for new warships, and Italian firms took the larger share from British ones. Italian businesses earned £2,918,900 compared to British sales worth £1,615,900. However, the three Italian cruisers *Buenos Aires, Garibaldi,* and *San Martin* that comprised the backbone of the Argentina's naval power did carry guns of Armstrong's latest construction.[47]

To pay for the arms race both countries employed foreign loans. In general, Argentina and Chile did not use direct taxation to generate revenue. Instead, indirect taxes on trade served as the principal source for government funds. Argentina, Brazil, and Chile were generally able to issue foreign bonds regularly after 1870, and their bond issues usually were well subscribed by foreigners, especially by British investors.[48] Back in the 1870s Argentina and Chile had comparable foreign trade and government revenues, but by 1900 Argentina's foreign trade had become almost three times as large as Chile's, with a corresponding increase in revenue 1.5 times more than Chile's. Argentina main exports included livestock and grain, while for Chile mineral exports dominated, especially saltpeter from the territories conquered from Peru in the War of the Pacific. Saltpeter generated 60 percent of Chile's revenues in the 1890s. Also, immigration had tipped the balance in favor of Argentina. In 1881 Argentina's population numbered 1.8 million, Chile's 1.6 million. By 1895 Argentina had 4 million to Chile's 2.7 million.

Both states made frequent recourse to foreign loans. Argentina had borrowed 710 million gold pesos in 1886–1890, and the public debt climbed from 80 million gold pesos in 1875 to 393 million by 1894. By 1896 Argentina's debt had reached 421 million gold pesos. In 1900 Argentina consolidated its debt into a conversion loan in state bonds, and in the process Argentina adding 49.2 million gold pesos into the nominal value to be used for armaments purchases. Chile suffered from a depression in the export price of saltpeter and turned more directly to foreign loans to sustain its military and naval orders. In 1895 Chile had borrowed £2 million from Rothschilds in London without which it could not have purchased the Krupp guns. The loan was so vital that Chile willingly paid Krupp's highest bid price for the guns to secure the funds. However, Chile had overextended its credit, and by 1898 German and English banks had mutually agreed not to grant any more loans to Chile until the tensions with Argentina were resolved.[49]

A comparison of military spending in the budgets of the two countries reveals the interactive nature and escalating pace of the arms race. Chile had achieved superior strength on victory in the War of the Pacific. Balmaceda augmented that power even more with military and naval expansion that began in 1889. In the post-1891 years Chile's military spending continued its upward trajectory. From 1892 to 1894 annual naval and military spending increased to an average of 24.27 percent (up from 20.29). Even more telling, for the first time in 1893 Chile's naval budget exceeded the army's budget 17.81 percent to 11.71 percent (Table 2). Likewise, in 1894 Chile's budget again gave greater resources to the navy over the army. Meanwhile, Argentina initiated its challenge to Chile by doubling its military allocations from 22 percent in 1891 to a whopping 40 percent of the annual budget in 1892, coinciding with the placement of the Mauser order.

Taking into consideration the larger size of Argentina's budget, Chile had every reason to worry. From 1892 to 1902 Argentina outspent Chile on military outlays every year. In 1894, the Argentine government requested an extraordinary military credit in a special congressional session. The government imposed secrecy on the press and threatened to shut down any newspaper that dared to report the details. Also at that time Chilean reconnoitering parties were violating Argentina's border. In January 1895 both chambers of the Argentine legislature voted £2 million (40 million marks) for armaments, of which 6 million to 8 million marks went for field artillery and engineers. By March 1895 Chile had already spent £289,300 (5.8 million marks) on armaments. The peak of the race occurred in 1896–1898 as both sides volleyed increasing naval orders at each other. Argentina overtook Chile and gained naval superiority with the purchase of the Italian-made *Garibaldi*-class cruisers. Chile, having now lost its former naval supremacy in the Southern Cone, bought the 4,500-ton protected cruiser *Chacabuco*, three destroyers, and a pair of transports. Argentina immediately responded with the purchase of two additional *Garibaldi*-class armored cruisers that were more powerful than the previous four. In desperation, Chile raided funds from its gold conversion reserve and ordered a pair of 12,000-ton battleships, thus prompting Argentina to order two 15,000-ton battleships from Italian yards and six destroyers originally laid for the Italian navy.[50]

After a decade of galloping procurements, in 1902 through British mediation the race came to a halt. Argentine President Julio Argentino Roca reported to Congress on May 8, "Happily, it appears that a better and more cordial understanding will be established in our relations with the Chilean Republic, negotiations having been opened in Santiago through the friendly mediation of the British Government for the rational limitation of the armaments which are pressing on both countries, with great prejudice to their credit and well-being."[51] The U.S. ambassador in Buenos Aires, William P. Lord, explained that the arms race had ended because of the effect it had on the creditworthiness and material well-being of the two countries. Significantly, large British interests in both

Table 2   Argentine and Chilean military expenditure as percent of budget (in millions $ US)

Year	Argentina		Chile		Chile	
	$	(%)	$	(%)	% Army	% Navy
1890	7.3	(20)	6.9	(18.42)		
1891	7.1	(22)	20.4	(50.89)		
1892	12.9	(40)	5.5	(19.06)	10.87	8.19
1893	14.3	(39)	6.4	(29.52)	11.71	17.81
1894	14.3	(37)	5.0	(24.22)	10.44	13.78
1895	14.9	(32)	11.0	(33.90)	25.15	8.75
1896	33.8	(45)	17.7	(42.0)	15.00	27.00
1897	28.7	(49)	8.4	(27.15)	13.59	13.56
1898	37.3	(32)	7.8	(43.47)	29.90	13.57
1899	28.7	(40)	7.3	(23.53)	14.22	9.31
1900	17.1	(26)	10.6	(18.46)	10.57	7.89
1901	16.5	(26)	10.5	(23.54)		
1902	23.8	(26)	19.2	(43.49)		
Total	258		138			

Sources: George v. Rauch, Conflict in the Southern Cone: The Argentine Military and the Boundary Dispute with Chile, 1870–1902 (Westport, CT: Praeger, 1999), 157; William F. Sater, Chile and the War of the Pacific (Lincoln: University of Nebraska Press, 1986), 274.

countries anxiously hoped for a peaceful solution, believing that it
would relieve the business depression and improve the financial condition of both countries. Lord wrote:

> Both countries are largely in debt and confronted with a deficit. Both
> have appropriated their conversion funds, which had been set apart for
> a specific purpose and which, it would seem, should have been pre-
> served as inviolable. Neither is able to make a foreign loan without pay-
> ing a high rate of interest and giving guarantee to meet the additional
> expenses which their war policy is incurring . . . With this condition of
> affairs confronting them, Argentina and Chile fully realize that their
> war policy is fraught with ruinous consequences to their credit, their
> Governments and their people, and that some means should be devised
> of stopping costly expenditures.[52]

Under the terms of the 1902 Naval Disarmament Treaty between
Chile and Argentina, both were to refrain from taking over the vessels of
war under construction for them. Each government agreed to reduce its
fleet. Article II bound both states to not add to their naval armaments for
five years without notice, nor were they to transfer any vessel to a country that had questions pending with one or the other. Under Article IV,
to establish the just balance between the two fleets, the republic of Chile
was to disarm the battleship *Capitan Prat,* and the Argentine republic
was to disarm its battleships *Garibaldi* and *Pueyrredon.* Britain served as
the arbiter of the treaty.[53]

Within Chilean naval circles there was interest in retaining the two
warships on order in Britain. As a possible compromise it was suggested
that the vessels ordered in England by Chile and those ordered by Argentina in Italy should be divided between the two countries, Chile taking an English vessel and an Italian one and Argentina doing the same.
The English vessels cost about £2.1 million. Private offers had been
made to Chile for £1.5 million, but Chile was not disposed to make such
a great sacrifice.[54]

East Asia

A very different sort of arms race played out in East Asia. The Japanese
methodically pursued their arms buildup as part of a coherent national
plan to join the ranks of the Western imperialist powers. As a result, the
arms race in East Asia was heavily one-sided. Japan crossed the threshold

from defensive to offensive armaments policy in 1886 when internal government discussions drew links between expanded military power and Japanese imperialist expansion onto the Asian mainland. The Japanese first set their sights on overcoming China, and after 1895 accelerated to fight Russia. The Japanese imported arms with specific adversaries in mind and as part of a comprehensive modernization and industrialization strategy. Like many other countries, Japan looked to British-built warships and Krupp artillery.

In 1886 Katsura Taro and Kawakami Sokoru, deputy chief of the General Staff, issued a joint memorandum laying out a more ambitious purpose for Japanese armament. They wrote, "Nations maintain an army for two reasons. First, to defend themselves against enemy attack or to preserve their independence. The armies of most second-class European nations are of this kind. Second, to display the nation's power, resorting to arms when necessary to execute national policy, as in the case of first-class European powers. Japan's aim in maintaining armed forces is not that of the second-class nations but that of the first-class powers."[55] As reflected by this memorandum, the vision and purpose for Japan's armed forces went beyond mere national defense. By the 1880s as Western imperialism carved up more of the globe, Foreign Minister Inoue Kaoru wrote that Japan should move quickly to establish its own empire on the edge of Asia before it was too late.[56] From this point Japanese leaders embarked on a pattern of armaments expansion to surpass China.

In a strategy reminiscent of the Russians, the Japanese demonstrated a preference for acquiring technology and know-how through licensing agreements. As part of the strategy the licensing arrangements always included the purchase of one or two foreign-built pieces, which were then studied by Japanese engineers. The Japanese were not infallible in their choices, however, as the example of Italian artillery makes plain. In 1885 Japan contracted with the Italian army to license its methods for producing bronze artillery. In 1887 the Osaka arsenal dutifully domesticated the processes for the 7cm gun, and by 1889 the entire Japanese infantry had bronze artillery for its armament just as the rest of the industrializing countries had made the transition to steel artillery.[57]

During the 1880s Japan developed domestic rifle production and even its own model design. In 1883 the arsenal held 2 million rifles of Snider and other patterns, all breechloaders, and a large supply of cartridges. The arsenal's small arms factory employed 700 workers, and turned out

10 rifles and 1,500 rounds daily of the Murata pattern breechloader rifle. The Japanese made gunpowder at Itabashi in the north and Meguro on south side of Tokyo and also at Mayebashi and near Takasaki. Infantry carried Enfield Snider rifles, but 100,000 mountain rifles were under preparation. In 1886 Japan adopted the Murata, produced it at the Tokyo arsenal as its standard, and the assortment of foreign firearms receded into the background. In a further sign of Japanese self-reliance, by 1898 the Murata rifle superseded the Martini-Henry.[58]

Japan, like many other countries, embraced the German model for its army in the 1880s, and that choice also brought a preference for importing Krupp artillery pieces. In the early 1880s Japanese artillery still employed rifled four- and twelve-pounders and bronze four-pounders on the former French system of muzzleloaders. Although the French firms Schneider and La Seyne together sold some 98 guns to Japan in 1884–1893, these figures paled next to Krupp's. Japan had bought only 47 Krupp guns in 1865–1874, but by 1883 Japanese facilities could make projectiles for Krupp guns. By 1886 the Japanese army had purchased 215 more Krupps, and from 1887 to 1899 Japanese acquisition of Krupp artillery doubled to 489.[59]

Although Japanese naval yards still lacked capacity to build large warships, the navy created the foundations for a naval armaments industry by setting up naval arsenals at Tsukiji in Tokyo and at Yokosuka. By the mid-1880s Japan began to phase out use of the sail and turned exclusively to steam-powered vessels. In 1882 the Japanese began phasing out wooden warship construction. They again turned to foreign assistance, this time British, to build iron warships. After importing machinery and hiring British skilled workers, Yokosuka arsenal was prepared to manufacture a few iron warships in 1884. Still, imported ships continued to be most important. In the mid-1880s the British built the second-class steel-decked cruisers *Naniwa* and *Takachiho*. These cruisers had displacement of 3,650 tons and were armed with 10.2-inch Krupps.[60]

The French played an increasing role when Emile Bertin took over management at the Yokosuka yard in 1886. The Japanese, following the advice of Bertin to abandon armor-clad construction and expend all the money they could to obtain cruisers and torpedo boats, purchased three powerful cruisers, two Armstrong and one French. The Japanese cruiser *Unebi-kan* had been built at Havre, but was lost at sea when its extremely heavy armament caused it to sink. The Japanese had bought *Unebi-kan* because they had wanted to have a vessel that could engage the formidable Chinese ironclads built in Germany. La Seyne was building two protected cruisers for Japan, *Itsukushima* and *Matsushima*, in 1888. These were all-steel steam warships with displacement of 4,000 tons and armed with 6-inch Canet guns and 4.7-inch quick-fire guns. French suppliers also received the majority of orders for torpedo boats as Japan bought ten torpedo boats from Creusot in 1887–1892. Thus, out of the twenty-one torpedo boats imported from Western countries, the majority were French (thirteen), five were British-made, and three German. Bertin also oversaw the Japanese construction of their first steel-hulled ships, *Atago* and *Takao*.[61]

In the first half of the 1890s up to the Sino-Japanese War (1894–1895), Japan returned again to British shipbuilding yards for its new cruisers. In 1892 Japan purchased the 4,150-ton *Yoshino* from Armstrong at Elswick. The ship carried twelve-inch quick-fire guns, and at the time it was the fastest cruiser in the world. Out of the seven imported cruisers in the Japanese fleet at the time of the war with China, five came from British yards and only two were French. The Imperial Navy also included seven smaller cruisers built in Japan.[62]

Legislative support legitimated arms expansion. The promulgation of the Meiji Constitution in 1889 allowed for the Diet to have some authority in budgetary matters. From the outset the sharpest conflicts between the Diet and government centered on the budget. New cabinets demanded and received tax increases for the armed forces and armaments, since military spending was the fastest-growing portion of the budget. In 1893, when the Diet was balking at appropriating funds for naval expansion, the emperor lowered the salaries of all officials by 10 percent for six years and pledged an annual contribution from the palace in the amount of 100,000 yen. The imperial example of austerity and discipline had the desired effect, and the Diet approved the funding increases. The Diet also proved an ardent supporter of the wars with China and Russia by voting money for both conflicts without hesitation. In those cases the Diet reflected general popular opinion in Japan.[63]

Japanese desire to dominate Korea drove the conflict with China, but the arms competition between China and Japan shaped the forces engaged. On July 12, 1894, the Japanese government resolved to start a war with China and instructed the Japanese minister in Seoul to "use any pretext available." Both sides had imported their main warships from Europe. China possessed larger ships, and its navy counted more tonnage. However, the lack of unity, which had posed problems in the procurement process, prevented the Chinese fleets from acting in a co-ordinated manner. Furthermore, China had not added any new ships since 1888. In contrast, Japan had bought nine fast cruisers since 1889. Japanese forces had uniform armament, and the Japanese fleet deployed its rapid-fire guns with devastating effect.[64]

Japan's victory over China in the Sino-Japanese War (1894–1895) opened the way for Japanese imperial expansion onto the Asian mainland, but Russian diplomatic intervention with support from France and Germany compelled Japan to relinquish much of its gains. Russia now emerged as the chief adversary, and Japan's leaders prepared to build up their military and naval forces to overcome the European Great Power. Using the Chinese indemnity (£37,836,127) as a preliminary step, Japan transferred £3,062,500 to the Naval Maintenance Fund.[65] Once again, Japan would find much aid through the international arms trade.

Although Japan strove to domesticate artillery production, Japan's arsenals could not satisfy the demand, so European guns were essential. In 1897 trials of various foreign quick-firing field and mountain guns took place, including those of Krupp, Hotchkiss, Canet, and Elswick. Great secrecy was observed, but the Japanese authorities decided to adopt and manufacture quick-fire guns of their own pattern, paying royalties to the foreign houses for various fittings.[66] Schneider-Creusot produced some field guns for Japan according to Japanese design. The French considered the design faulty and expected that it would be obsolete in a few years. Schneider's engineers found the Japanese model overly complicated and difficult to make, and the firm's experts did not think highly of them. Nevertheless Schneider-Havre accepted the business and in July 1900 was making 110 mountain guns for Japan on Japanese design. Schneider also made sixty quick-fire guns for Japanese coastal defense.[67] Typically, Krupp benefited most from Japanese purchases. In preparation for war with Russia, in 1900–1903 Japan ordered a staggering 2,136 Krupp guns, making Japan Krupp's single biggest customer in the world in the first half of the decade.[68]

Experience in the Sino-Japanese War taught the Japanese that they required more heavily armored and faster battleships that would be capable of a higher rate of fire while providing heavier armament. The Ten-Year Naval Expansion program passed in 1896 called for acquiring four battleships, six armored cruisers, twenty-three torpedo boat destroyers, and sixty-three torpedo boats. The bulk of the orders went abroad since Japan's own industrial resources at the end of the 1890s were not yet up to the task of constructing a main battle force of armored warships. Japan's plants were still obtaining newly integrated steel plants, furnaces, and rolling equipment. Consequently, roughly 90 percent of the 234,000 tons of new warship construction had to be imported. These new foreign-built ships ultimately made up 70 percent of Japan's battle fleet in the war with Russia.

British firms garnered the majority of these major contracts. The two Japanese initial battleships, *Fuji* (Thames Iron Works) and *Yashima* (Armstrong), delivered in 1897, were not part of the 1896 extension plan, but they immediately became the largest and most powerful vessels in the Japanese fleet at time. All four of the new battleships demanded by the Ten-Year Plan came from British yards. Thames Iron Works built the *Shikishima*, Armstrong provided the *Hatsuse, Asahi* came from John Brown, and the *Mikasa* was built by Vickers. Among the six first-class armored cruisers, Armstrong sold three (*Asama, Tokiwa, Idzumo*). French and German shipbuilders sold one each: St. Nazaire built the *Adzuma*, and Vulcan made the *Yakumo*. The last first-class cruiser of the six was Japanese-built. The Japanese also imported three second-class cruisers in 1897–1898. Once again Armstrong claimed a share, building the *Takasago*. The other two vessels were made in the United States. Cramps in Philadelphia launched the *Kasagi,* and Union Iron Works in San Francisco built the *Chitose.*[69] In 1902, twenty-two first-class torpedo boats were under construction abroad: five Yarrow, six Normand, and eleven Schichau.[70]

To a lesser degree domestic production also contributed to Japanese naval expansion. In 1897 the Tokyo arsenal turned out its own torpedoes based on Whitehead designs, and Japanese yards built five of eight protected cruisers in the period 1894–1904 on Japanese design, using Japanese materials but British imported guns. Starting in 1901 Japan produced its first armor plate. Lastly, by 1902 thirty-nine destroyers were under construction in Japan.[71]

As a reflection of the desire to standardize armament the Japanese intended to employ only Armstrong guns for their naval armament. Krupp and Canet guns were phased out starting in 1898. Also, the government erected a gun factory in Kure Dockyard with English equipment to make guns up to 24cm.[72] The Japanese selection of Armstrong guns reflected the most advanced technological system in the era. By the 1890s Armstrong and Whitworth had pioneered quicker and more powerful guns with increased muzzle velocity and using smokeless powder.[73]

Given the major augmentation in Japanese military and naval power it is perhaps surprising that Russia did not respond in kind. Russia opted not to join the race with Japan, most likely because St. Petersburg felt so far ahead. By itself Russia's Far Eastern fleet possessed more capital ships than the entire Japanese navy. By 1898 the Russian navy held third place among world navies in terms of number of warships with 107, behind Britain (355) and France (204), but ahead of Germany (77). The Naval Ministry cast a wary gaze at Japan and lobbied for increased ship construction for Russia's Pacific fleet. The navy's proposal was opposed by the War Ministry and the ministry of finance. Sergei Witte, minister of finance, argued that Russia as a land power should not endeavor to match Japan in naval forces, but rather should rely on the future completion of the Trans-Siberian Railroad as the principal means to bring Russian ground forces to bear on Japan if necessary.[74]

Russian budgetary expenditure confirmed that Russia did not engage in an all-out arms race against Japan. In fact, since the 1860s the army and navy had received a decreasing share of the national budget, and that trend had continued up to the Russo-Japanese War. From 1898 to 1903, the army's budgetary allocations on armaments increased nearly £4.5 million, or approximately 8.5 percent of the total increase in all national expenditure for that five-year term.[75] Nevertheless, on average, total military expenditure (navy and army) as a percentage of the whole budget declined from 28.1 percent to 22.4 percent between 1890 and 1904. The naval expenditure increased slightly in that same period, from 4.3 percent to a high of 5.5 percent. Overall, though, Russia easily outspent Japan in naval expenditures (Table 3).

The Naval Disarmament Agreement between Chile and Argentina placed some ready-built warships up for grabs between Russia and Japan. Russia had offered £1.87 million for Chilean warships being constructed in Britain. Unless Japan met the Russian offer, Russia could buy and thereby obtain preponderance in Far Eastern waters. The sudden intervention of Russia raised British fears. Britain made every effort to postpone Russia's purchase to give Japan the chance to reconsider the matter. The last offer from Japan stood at £1.55 million.[76] The British naval attaché in St. Petersburg reported that the Russian Admiralty strongly opposed buying these completed ships abroad, owing to the differences in armament.

It would also be impossible to rearm them before the spring of 1904. The Russian interest seemed calculated to prevent the Japanese from buying the vessels.[77]

The Japanese naval attaché in London strongly urged the purchase of the South American ships, which were considered a great bargain. The Japanese Admiralty looked favorably on the acquisition but the money difficulty continued, and the question of their purchase had not yet been considered by the Cabinet.[78] The Japanese government had offered £1.6 million for the ships. This amount could have been obtained from the Naval Reserve Fund, the interest on which (about 6 million yen) went toward the upkeep and repair of the navy. Under ordinary circumstances this fund could not be touched without the consent of the Diet, but as the Diet was not then sitting, the necessary authority to utilize this fund could at that time have been obtained by an imperial ordinance. As the Diet was scheduled to meet in a few days, Cabinet members were unwilling to use an imperial ordinance, believing that the Diet would strongly resent any such action. The Cabinet carefully considered the offer and decided not to buy the ships. The Japanese government

Table 3   Total Russian and Japanese annual naval expenditure (including construction and armaments), 1889–1903 (in millions £)

Year	Russia	Japan
1889	4.078	1.515
1890	4.268	1.142
1891	4.946	1.544
1892	5.218	1.332
1893	5.484	1.080
1894	5.511	1.110
1895	5.713	1.464
1896	5.953	2.167
1897	8.290	8.667
1898	9.028	7.500
1899	10.607	5.076
1900	10.963	4.178
1901	11.660	3.712
1902	10.446	2.899
1903	11.094	2.848

Sources: ADM 231/38, Rept. 700, Foreign Naval Progress and Estimates, 1903; ADM 231/32, Naval Estimates, 1900; ADM 231/44, Naval Progress and Estimates, 1905.

telegraphed Count Hayashi Tadasu in London to that effect on November 26, 1903. On December 2 Claude MacDonald had a long conversation with Count Komura Jutaro, who explained that the problem was financial. He added that it would have been overcome had not the sitting of the Diet been so imminent. Count Komura said that the Cabinet had so many thorny questions in front of them, and expected such troublous times during the next session, that members were unwilling to add to their number. As to the Chilean ships themselves, the main reason why the Japanese government wanted to buy them was to prevent the Russian government from doing so. The ships, he understood, were excellent, but the guns would not take Japanese ammunition. This difficulty could have been overcome, but the sudden rise in price combined with the difficulty of getting the money at such short notice convinced the government not to buy. According to Japanese information, Chile had approached Russia, not vice versa. Komura considered that Chile, to get a good price for their ships, had introduced competition in the form of the Russian government.[79]

With Japan financially unable and politically unwilling to purchase the Chilean ships, Britain stepped forward to aid its ally. To assist Britain's own naval construction program, and to "obviate a disturbance of naval power to the disadvantage of our Japanese allies," the British government decided to buy the ships in question. The British ambassador in Tokyo pointed out to the Japanese "that HMG had taken this step in order to strengthen the hands of the Japanese Government."[80] The Japanese minister of Foreign Affairs expressed appreciation for Britain's action.[81]

The Japanese minister called on MacDonald on December 17, 1903, and asked whether Britain would be prepared to let the Japanese government purchase from them the two warships recently acquired from Chile. He explained that previously the Japanese government had felt unable to offer the demand price because it lacked the Diet's consent. With the Diet dissolved, there was no doubt that the money would be voted on the reassembling of the parliament. MacDonald replied that the two ships were now part of the British fleet since the Japanese said they did not really want them. "I reminded Hayashi that the two cruisers built in Italy for the Argentine government were still in the market. I was under the impression that although less valuable than the Chilean ships they were nevertheless powerful and valuable vessels." Moreover, "transfer of the ships from the British to the Japanese Navy at the present moment would almost amount to an act of open hostility to Russia."[82] MacDonald ascertained that the proposal to buy the ships from Britain had emanated from Count Katsura Taro, the prime minister, and a member of the Genro, the remainder of the Council agreeing that it might be made. The minister of war thought Admiral Yamamoto Gombei had been very unwise in not seizing the opportunity to buy the ships when they were first offered.[83]

The Japanese took MacDonald's hint. They bought the Argentine ships *Moreno* and *Rivadivia*. Japan paid 10 percent down and required that the ships sail from Genoa with full munitions. An agent from Armstrong inquired whether the British consul in Genoa would register the ships as British merchant vessels and allow them to sail under British flag. The consul strenuously refused. The ships made it to Japan before the start of the war.[84]

The South American arms race fit the classic action-reaction model. Chilean purchases stimulated an Argentine response, which then spurred another Chilean action, and so on until an escalating spiral ensued. The race had originated from Chilean ambitions flowing from Santiago's newfound prowess after victory in the War of the Pacific. Argentina joined the race because Buenos Aires had coveted the position of premiere South American power as its own after having caught up to and surpassed Brazil in the preceding decade. Nevertheless, German private business interests, with the connivance of German diplomats and army officers, served as a catalyst to accelerate the rate and size of the spiral by consciously and covertly fanning the flames to increase artillery and rifle sales. Krupp and Loewe pursued the South American sales aggressively in the 1890s because their established markets in Eastern Europe were declining. The Russians bought nothing, the Balkan states mostly bought Austrian rifles, and the Turks did not buy as many armaments as they had in the previous era. To compensate for the shrinking market size, the strategy of Krupp and Loewe centered on maximizing buyer demand as much as possible where the demand clearly existed.

The modesty of Krupp's naval orders highlights an important point, namely that the Germans had not achieved influence over all of Chile's armed forces, but rather only over the elites in the Chilean War Ministry and army. Balmaceda's naval flirtation with the French in 1889–1890 ended abruptly with Chile's civil war. After the 1891 revolution the Chilean navy turned its back on French suppliers, but it did not favor Germany. Consideration of the naval side helps balance out our understanding of the role of the arms trade in the nascent South American arms race. If we take the military side in isolation we could easily misread the dynamic as entirely generated by Krupp and the Germans. Yet the naval purchases figured far more significantly in the Chilean-Argentine rivalry, and in that sphere the Germans remained bit players. Also, as noted above, Chile had initiated its naval expansion under Balmaceda in 1889, prior to Körner's rise to influence.

The Japanese arms buildup raises interesting insights about the interplay of international and domestic forces in shaping an arms race. Witnessing the scramble of Western imperialism in Asia, the Japanese leadership concluded that Japan needed to enter the fray and secure its piece of China before it was too late. While participating in imperialism on the Asian mainland, the Japanese imported the weapons and the mindset of imperialism. Yet without firm domestic support for the expansionist agenda Japan would not have been able to pay for the imports. If Japan had relied on foreign loans, as the South American countries did, external financial interests might have applied leverage to inhibit Japanese expansion. Alternatively, if Russia had chosen to race against Japan it seems very likely that additional Russian naval units deployed more quickly to the Pacific would have deterred the Japanese from launching the war as they did. Such a more powerful response by Russia, however, might have expanded the conflict by drawing in active British involvement. Obviously such counterfactual arguments cannot be proven. Without a doubt, though, Japan could not have ascended into the ranks of the Great Powers as it did without resorting to the global arms trade, but the Japanese harnessed the business for their own ends rather than allowing themselves to be manipulated by sales agents.

6

The Dreadnought Races

With the launching of HMS *Dreadnought* in 1906 the British redefined the naval trade in the years up to World War I. By 1914 the British firms Vickers and Armstrong, working in tandem, unquestionably emerged supreme in winning contracts. In the end, Vickers won the lion's share of these major contracts. The two shipbuilders won £35.87 million in sales between 1900 and 1914, or roughly 63 percent of the total business in the combined markets of South America, East Asia, Turkey, and Greece. French sales ranked a distant second with £5.3 million; Italian and U.S. firms sold roughly £5 million each; the Germans, £4.3 million; and the Austrians a little over £1 million.[1]

From 1906 to 1914, besides the more famous Anglo-German naval race, regional naval races occurred in South America and the eastern Mediterranean. Unlike the Anglo-German race, these lesser naval competitions relied almost entirely on imported warships. In South America the Brazilians, having been eclipsed by Argentina, led the charge with the goal of reasserting Brazilian weight as the regional power. The Brazilian-Argentine race differed from the previous Chilean-Argentine one in that the Brazilians competed solely in the naval sphere. This was a race about who had the most and biggest dreadnoughts. Not only did the race lack a broader strategic purpose; it also was disconnected from a larger defense context since the artillery and rifle purchases did not exhibit the same move-countermove dynamic that the ship purchases clearly did. Brazilian foreign minister José Maria da Silva Paranhos Júnior, the baron Rio Branco, acted as the driving hand, and his death brought an end to the Brazilian buildup. The Greco-Turkish naval race was fraught with far more peril. In seeking to establish modern domestic naval industry the Young Turks, a group of revolutionaries, indicated their aspiration to cast off the atrophy under Sultan Abdulhamid II and revitalize the Ottoman Empire as a dominant regional power. Meanwhile, Greece pursued naval expansion with the conscious aim of competing with the Turks. The Greeks planned to take territories from the Turks, including some of the islands, and they were hurrying to gain the advantage to strike before Turkish moves could thwart them. The two came to blows during the First Balkan War in 1912–1913 to the advantage of the Greeks.

After 1908 prospects for extremely large naval orders for Eastern Europe increased sharply (Table 4). Russia's Pacific and Baltic fleets were destroyed during the war with Japan, and significant orders for warships appeared on the horizon now that the tsarist government had the financial means to pay for the rebuilding programs. In addition to purchasing ships outright, both the Russians and the Turks desired to have a domestic capability to build dreadnoughts. The geographic isolation of Russia's Black Sea fleet led to the establishment of dreadnought construction facilities on that seacoast, part of the competition between the Russians and the Turks. On the eve of World War I, Vickers had established joint ventures to provide domestic dreadnought production facilities for the Russians and the Turks on opposite sides of the Black Sea.

South America took part in a second naval race, this time spearheaded by Brazil. The main issue in the foreign relations of the South American states revolved around the balance of power among Argentina, Brazil, and Chile. As early as 1902 Argentina stood on the brink of war with Chile, and the possibility existed for Brazil to have taken that opportunity to cooperate with Chile to prevent Argentina from dominating the continent. Brazil could have counted on assistance from Uruguay, which had a standing grievance with Argentina over the River Plate. Bolivia might have joined, too, but Chile would have found Peru as a threat to its rear. At the moment Chile had excellent relations with Brazil. Argentina wanted to be left in peace to enjoy its agricultural and commercial prosperity. However, any change to the status quo by Brazil would call forth a response from Argentina. By 1905 Chile considered all possible problems with Argentina at an end. Both governments considered the disarmament treaty unnecessary and undesirable since it prevented Argentina from responding to the extensive naval construction of Brazil, and Chile could not adopt prudent

Table 4   Cruiser, battleship, armored vessel, and ironclad exports delivered by firm and region, 1863–1914

Firm	Asia	South America	Eastern Europe	Total
Laird (BR)	—	6	—	6
Samuda (BR)	1	3	1	5
Armstrong (BR)	13	15	3	31
Vickers (BR)	3	1	1	5
La Seyne (FR)	2	6	6	14
St. Nazaire (FR)	—	—	1	1
Vulcan (GER)	9	—	1	10
Schichau (GER)	—	—	1	1
Ansaldo (IT)	2	2	—	4
Orlando (IT)	—	2	1	3
Danubius (AUS)	1	—	—	1
New York SB	—	1	—	1
Fore River (US)	—	1	—	1
Total	31	37	15	83

Sources: ADM 231/10, Rept. 119, Shipbuilding Policy of Foreign Nations 1886; FO 6/490, Haggard (Buenos Aires) to Lansdowne, 13 May 1905, Enclosure, Juan A. Martin, Memoria del Departamento de Marina 1904–1905; Conway's All the World's Fighting Ships, 1860–1905 (New York: Naval Institute Press, 1976); Conway's All the World's Fighting Ships, 1906–1921 (London: Conway's Maritime Press, 1985).

measures regarding four warships ordered in Italy by Peru. Chile wanted to unload its old warships and remodel with new ones.[2]

Rear Admiral Alexandria de Alençar, Brazil's naval minister, was anxious to reform the navy. He hoped to make Brazil the dominant naval power of South America. Under Admiral Julio de Noronha, naval minister in 1903 under President Rodriguez Alves, the Brazilian government had decided how to increase its navy.[3] Brazilian Naval Commission officers formulated the building program in 1904. If not for the naval revolt, the supposed monarchical sympathies of the officers, and want of money the government might have begun earlier. By 1906 the Brazilian navy was inferior to the Argentine in number and type of ships. The naval minister of the outgoing Brazilian government had placed orders for new vessels of only 13,000 tons each, and the Brazilian Congress authorized £1,685,820 out of a total of £4,214,550 for construction of new warships to be supplied by foreign firms in 1906 (Table 5).[4]

Table 5   Torpedo boat and gun boat exports delivered by firm and region, 1865–1914

Firm	South Asia	South America	Eastern Europe	Total
Yarrow (BR)	15	60	12	87
Thornycroft (BR)	—	10	—	10
Laird (BR)	10	5	5	20
Armstrong (BR)	12	2	—	14
Thames (BR)	—	—	8	8
La Seyne (FR)	—	—	5	5
Normand (FR)	31	—	—	31
Schneider (FR)	—	—	14	14
Vulcan (GER)	18	—	18	36
Schichau (GER)	19	7	59	85
Germania (GER)	1	2	2	5
Cramps (US)	—	—	30	30
Kawasaki (JA)	15	—	—	15
Ansaldo (IT)	—	—	11	11
Total	121	86	164	371

Sources: ADM 231/10, Rept.119, Shipbuilding Policy of Foreign Nations 1886; *100 Jahre Schichau, 1837–1937* (Berlin: VDI-Verlag, 1937), 186–87; *Conway's All the World's Fighting Ships, 1860–1905* (New York: Naval Institute Press, 1976); *Conway's All the World's Fighting Ships, 1906–1921* (London: Conway's Maritime Press, 1985).

Brazil had remained aloof from the neighboring arms race in the 1890s, largely because domestic political developments absorbed the attention of the armed forces. The Brazilian army had overthrown the monarchy in 1889 and installed a republic. After having helped establish a constitutional order the army lost some of its credibility because of internal political battles. Also, as author Frederick Nunn states, "Unlike Chile, independence did not provide the army with a glorious tradition or the makings of one to justify its importance or existence."[5] During the civil war, most of Brazil's fleet and the arsenals were either destroyed or damaged. Subsequent attempts to rebuild the navy were blocked by the Brazilian army because of lack of faith in the navy's loyalty.[6]

During the 1890s the Brazilian navy did not possess a single first-class warship. Brazil bought lesser warships from British, French, and German firms. Armstrong supplied two second-class and one third-class armored coast defense ships and one torpedo vessel. La Seyne provided three new armored coastal ships and renovated an old battleship. For the smaller craft, Thornycroft and Schichau each sold three torpedo boats, and Germania sold two gunboats.[7]

On January 23, 1906, the Brazilian warship *Aqidaban* with several high officers on board was blown up. Two hundred men died, including three admirals. Despite this loss and the lack of trained officers, the Brazilian Admiralty decided to order three great battleships from Armstrong. As usual, the British ambassador found these large armaments acquisitions appalling even though a British firm was the beneficiary. W. Haggard viewed the entire battleship program as a waste of money, and it had, moreover, the effect of rendering the Argentine republic very suspicious of Brazilian motives. Haggard saw "an embodiment of national vanity, combined with personal motives of a pecuniary character" as the motives behind the orders.[8] On July 23, 1906, Brazil signed a contract with Armstrong for three vessels, each 13,000–14,000 tons, to be delivered within three years for £4.8 million. Brazil also had begun negotiations with Armstrong for construction of dock and arsenal facilities near Rio de Janeiro, or simply Rio, where these vessels could be repaired, and where Brazil might eventually construct future vessels domestically. In September 1906, J. Falkner and M. Hurzig from Armstrong signed a contract for three 13,000-ton battleships at £4.212 million. They also surveyed the Bay of Jacacangua for a proposed arsenal estimated at £2.225 million. However, some in Congress wanted the arsenal in Rio.[9]

Brazil's signing of the contract for construction of three battleships immediately caused concern in Argentina and Chile. The specifications for these warships required that they draw no more than 24 feet, an unusually small draught for vessels of 13,000 tons but the maximum draught possible on the River Plate, which raised alarm in Buenos Aires. In reaction the Argentine minister of foreign affairs informed Haggard that Argentina intended to order two battleships, and indicated that this order would undoubtedly go to England. According to Haggard, the Argentine government worried less about the practical efficacy of the Brazilian ships than that the Brazilian government might think that their possession of superior ships to those of Argentina might justify an act of aggression against their prosperous neighbors. Chile ineffectively joined Argentina in trying to induce Brazil not to purchase the battleships. They got no answer from Brazil. Argentina took the threat seriously, and the government prepared to spend £7 million on defense of the River Plate by land and sea. Chile too, moved to keep pace with Brazil and Argentina.[10]

Having just come through a naval arms race, Argentina did not want to endure another one. As a means to convey to Brazil the earnestness of its resolve, Argentina proposed to order its two battleships at once from England for £3 million, and use the remaining £4 million for land defenses over five years. Construction of the battleships was to begin and be paid for at once, so that Brazil would see that Argentina wasn't bluffing, according to Dr. Montes de Oca, Argentine minister of Foreign Affairs. In fact, the minister believed that Congress would vote the money, and his only fear was that it would insist on three ships instead of two. Argentina also intended to send a communication to Brazil to try to persuade Rio to get rid of its ships, in which case Buenos Aires would get rid of its. Argentina also sought to bring Chile into the arrangement, as Santiago would be obliged to follow suit in the purchase of ships. De Oca assessed the situation as preparing for war to preserve peace. The Argentine government seriously distrusted Brazil. In the course of an interview with Haggard, the Argentine foreign minister prefaced his statement with the remark, "These Brazilians have really gone mad." He then related that Brazil had been trying to purchase from Ecuador, at the price of £20 million, a piece of territory in the south of that country, which would give Brazil an outlet on the Pacific.[11]

Mutual suspicions between Argentina and Brazil led each to assume territorial expansion by the other. Argentina held the belief that Brazil dreamt of the acquisition of Uruguay and Paraguay, while Brazil believed the same of Argentina. The supposed Brazilian plan to acquire a Pacific strip from Ecuador was actually an Ecuadoran proposal to trade to Brazil a strip of territory actually in Peru's possession but claimed by Ecuador, in return for a slice of the Amazonian basin and $10 million. Brazil had ignored the proposal. Regarding the Brazilian navy, the civil war had either destroyed or greatly damaged the major portion of the fleet and the naval arsenals. Later, an attempt to rebuild the fleet was wholly abandoned owing to the strong opposition from the army. Brazil found itself with obsolete or out-of-repair ships and a rundown naval arsenal. When Argentina bought ships and built a great naval port, Brazil had not questioned the motives of Buenos Aires. Now Brazil felt bound, "in justice to her own place as a country and to her proper individual interests, to rebuild and refit what she has had and lost she cannot understand why Argentina should question her motives or presume that in this procedure on her part something lay hidden dangerous to Argentina."[12]

According to Argentina, Brazil's decision to guard its extensive coast line with three large battleships appeared unnecessary and dangerous. The Argentine press and the minister of foreign affairs opposed Argentina's own naval acquisition, but they could not undertake the responsibility of leaving the country unprepared in case of conflict. Windham Baring, the British ambassador, discerned internal instability as the driving force behind Brazil's naval policy. He reported, "Internal Brazilian politics seems to be in a very corrupt state, and it would not be impossible that if this republic (Brazil) were sufficiently armed it could attempt to wage an external warfare in order to prevent the dismemberment of the various Federal states, and it is also possible that the country it would have in view as its goal would be the Argentine, as there is very little love lost between these two nations, and the Brazilians are very jealous of the progress and prosperity of this country (Argentina)."[13]

The Brazilians were dismissive of Argentine objections to their naval expansion. The Brazilian press deemed it merely Argentine chauvinism, and refuted the Argentine charge that the shallow draught of the new ships (only twenty-four feet) proved Brazil's aggressive intentions. Articles pointed out that there were shallow estuaries other than the River Plate, and that it was expected that Brazil be able to send its ships up the Amazon. Brazilian officials stated publicly that they had no menacing attitude toward any neighbors. Privately, in diplomatic conversations, they told the British that they would like to feel more confident at sea because, in the frontier delimitation dispute with Bolivia, Argentina had been backing up the latter. The Brazilians felt convinced that this would never have been the case had they more sea power. "Their object therefore in ordering new ships would seem to be to insure their independence of action in questions of foreign affairs, and to give their foreign policy greater moral, and if needs be, effective support."[14]

Ernesto Tornquist, the Argentine financier who had been instrumental in preventing war with Chile in 1902, tried peaceful means for a solution. In Congress, Tornquist was fighting hard to prevent the Argentine ship purchase without again attempting to persuade Brazil to get rid of its ships. Tornquist proposed that the three Brazilian battleships be shared so that Argentina, Chile, and Brazil each had one. On September 28, 1906, Tornquist met with William J. Buchanan, a U.S. delegate to the Pan American meeting. Tornquist told Buchanan he had had a long conference with Argentine president Dr. Figuera Alcorta, and the president had agreed with the plan of exchanging men-of-war with Brazil and Chile in principle. Tornquist transmitted his suggestion to Brazil through

Buchanan, but Brazilian foreign minister Paranhos said the proposal was not feasible. Tornquist then suggested to the British that the way to keep Brazil in line would be to get Rothschild to cut off funding. Tornquist believed that Brazil's financial condition was deplorable, that Brazilians were living hand to mouth, and that the fear of getting no more advances from Rothschild ought to keep the country in line.[15]

In his presidential inaugural address in November 1906, Brazilian president Dr. Alfonso Penna presented the plans for new naval acquisitions as nothing more than a modernization of outdated forces. The incoming government decided to alter the naval program, and to order, instead, three battleships of the dreadnought type of 18,000 tons each.[16] The obsolescence of Brazil's navy in recent years justified the country's determination to replace them with the most modern systems. The president maintained that Brazil had not embarked on a policy of armed peace, "the curse and ruin of the nations constrained to it," but that "the preservation of the same naval and military forces we have maintained for years past unaltered, in spite of the great increase in population, and the increase of national and foreign commerce, is an eloquent testimony to the peaceableness of our intentions."[17] The naval minister and Paranhos, who was already directing foreign affairs, were responsible for the decision to build dreadnoughts against great opposition. In support of the president's policy in early December the Brazilian Congress authorized the government to make any desired alterations in the naval program, provided that the changes did not exceed the total amount of expenditure. The government opted to reduce the number of projected destroyers and to increase the tonnage of the battleships. Although the construction of the new battleships had barely begun the government had already selected and appointed their principal officers and sent them to England to supervise the progress of the work. The new dreadnoughts were designated *Rio de Janeiro, Minas Geraes,* and *San Paulo,* after the states that were parties to the coffee valorization scheme (a policy of state price supports).[18]

While aboard a steamship from Southampton to Buenos Aires in December 1906, Walter Townley, British ambassador to Argentina, traveled with Dr. José Carlos Rodriguez, director of *Jornal do Commercio* in Rio, and Emilio Mitre, director of *Nacion* of Buenos Aires. Each newspaperman expounded his views about the naval situation. Rodriguez defended the action of Brazil in desiring to increase its naval strength, urging that the country's extended coastline made it imperative. He said it was ridiculous of the people in Buenos Aires to imagine some menace to peace, or that it could be construed as a wish by Brazil to secure naval supremacy of South America. The Brazilian navy had deteriorated enormously. Now that Brazil was financially able to do so, it was duty bound in the interest of trade protection to reestablish the naval forces. Mitre took the view that Brazil's action would oblige Argentina to increase its fleet. He was scornful of Brazilians saying they were jealous of Argentine prosperity and prominence.[19]

Military and naval modernization figured prominently in President Penna's message to Congress in 1908. In his address on May 3, the president noted that with the coming construction of the smokeless powder factory and the installation of the proper plant in the arsenals to manufacture projectiles, Brazil would no longer be dependent on foreign markets for the purchase of ammunition. In naval matters Penna justified the modifications of the dreadnoughts to give greater efficiency to the warships then in construction. "These alterations were conceded by the contracting firm . . . with a reduction of more than 1 million pounds sterling, which was due to the adoption of the most modern type of battleship. . . . The Arsenal requires to be completely remodeled . . . to repair battleships of a modern type."[20]

German and French efforts to lobby the Brazilians for naval and gun orders took sharply divergent paths. For the Germans, the chosen approach aimed directly at the naval and war ministers. When the Brazilian naval program of 1904 had already been drawn up, but before the construction orders had been given, the Germans sent an admiral with an autographed letter from the kaiser to the naval minister, Admiral Alençar, asking that German interests not be neglected. The Germans found some success when in 1908 the admiral withdrew a contract with a British firm to supply electric plant for a marine arsenal (£3,000–£4,000). Alençar countermanded the order after German Minister von Reichenau had paid a long visit and had emerged with a promise that German industry would be encouraged to supply material for some or all the machines to be installed in the new electrical repair shop of Los Cobras Arsenal. Similarly, the invitation of the German kaiser to the Brazilian war minister for a visit to Germany had as its purpose an invitation to purchase guns and ships from German firms. This meeting had a very limited effect since the kaiser met with the war minister only once, and then only by a casual meeting at the maneuvers. Also, using the kaiser as a personal lobbyist had a serious drawback. Such a policy cheapened the kaiser and his ministers in the eyes of the Brazilian officials, who expected a comparably high level of attention when the next order came along.[21]

The French, in contrast, tried to employ their financial assets to secure the orders. Creusot and French banking connections protested the visit of Brazilian generals to Germany. To demonstrate their disapproval and the power of their financial opinion, the French warned that such conduct would prejudicially affect Brazilian credit. This point came across in an article in *Temps* September 24, 1908, in Paris on armaments of Brazil. The article gave voice to the French expectation that foreign countries that have financial relations with the Paris market should order their armaments from French firms. The article complained that Brazil had ordered three large battleships in England, and field guns, armored cupolas, and rifles in Germany. That French capital had been allowed to construct ports and railways in Brazil was deemed as a favor conferred by France, along with the fiscal concessions over the duty on coffee and the assistance given toward the valorization scheme. Whereas France had provided all these proofs of its friendship for Brazil, Brazil had repaid France's kindness by placing its naval orders in England and its gun orders in Germany. The article concluded that it was all well and good to talk about the intellectual relationship between France and Brazil, but Brazil should act by at least giving a small crumb to French enterprise.[22]

Paranhos's plan was to make Brazil strong, at least on paper, to impress the rest of South America, and to enable him to solve without conflict the numerous boundary questions. The baron considered even the mere notion that the battleships might not be retained as a blow to the prestige anticipated from their acquisition, and he denied any rumors that Brazil would transfer its dreadnoughts to another power. Paranhos expected that Argentina and Chile were each likely to build two battleships, but he seemed unconcerned that events in South America appeared to be leading to a state of armed peace. The Brazilian government was determined not to relinquish its ships, and officials confidently believed that they would be able to pay for them despite the strain caused by the acute depression. Everything depended on the recovery of the staple exports. Without it the population faced the prospect of more taxes to cover the extra naval armaments, in addition to the requirements for the foreign loans and an increasing foreign debt.[23]

The Argentine president was willing to withdraw the project from Congress if Brazil would sell its battleships under construction. Brazil had to decide immediately, as public opinion in Argentina would force Congress to pass the project.[24] On August 9, 1908, the Argentine president authorized British representative Colonel Sir Charles B. Evan-Smith to pledge the Argentine republic to relinquish all measures financial or otherwise for preparation of new armaments if Brazil would consent to sell its new ships. If necessary, Argentina would purchase one to facilitate agreement, though Argentina did not desire to do so. Unless Brazil concluded the agreement soon, the Argentine Congress would proceed to vote £10 million for armaments. Argentina had the pledge delivered through the British Foreign Office, in order that British diplomacy might be applied indirectly.[25] The Argentine minister of public works confidentially requested Sir Charles to get the following telegram published immediately in the *Times,* omitting the minister's name: "Senseless agitation regarding armaments in Brazil has produced its natural effect here. Argentine Government much against their will have had to give way to public opinion and have decided on immediate expenditure of 10 million pounds on military and naval armaments. This expenditure will be met entirely from ordinary revenue and will not necessitate any new loan or recourse to Conversion Fund formed by annual installments for eventual conversion of paper currency which has now reached 5 million pounds."[26] The British told the minister of public works of concern in London that publication of his message would have a prejudicial effect on Argentine credit.[27] The Argentine president wished the British to convey his intentions privately to Brazil merely to assure them of his willingness to avoid the naval race, but the "Argentine Government can and will make no official action towards Brazil, knowing this would be construed as weakness here and in Brazil. They are most unwilling to conclude War preparations Vote of 10 million pounds, foreseeing effect on market."[28]

Paranhos would not countenance even the suggestion that Brazil would sell its warships under construction as long as the possibility remained that Argentina might not respond with a similar program. In August 1908 a Brazilian official communiqué announced that £3 million, or about 60 percent of the total price, had already been paid to contractors, and that the annual expenses would now diminish as the ships were delivered. The announcement served as proof of Brazil's commitment. Francisco Herboso, the Chilean minister in Brazil, believed Brazil was earnest about its naval policy and was increasing the fleet quite independently of what might be done by others. Chile intended to build ships as soon as it had the money to spare, and it might be compelled to do so sooner if Argentina led the way. Paranhos intended to hold to his shipbuilding scheme. In this way Brazil could count on Chile's naval construction to match Argentina and thus use Chile as a counterweight to Argentina.[29]

The construction progress of the *Minas Geraes* at Elswick excited the keenest interest, and the newspapers devoted many columns of print to all the European news with regard to battleships and the building program. The *Jornal do Commercio* transcribed long articles from the English and French press.[30] Any action on part of the Penna administration to give away the ships would be a weapon in the hands of the president's opponents. In anticipation of congressional elections in January 1909, it would be imprudent to risk offending national feeling. The army and navy were unanimous in this respect, and "it is always a grave danger to a Brazilian president to have against him a popular sentiment which extends to the army."[31]

Having been unable to dissuade the Brazilian government from its naval expansion, Argentina moved ahead with its counter measures. The Argentine Chamber of Deputies voted 72–13 to pass the government's plan to increase the nation's armaments. The government expected the Senate would also vote approval, but for a lesser amount. General Roca could count on a majority in the Senate against heavy expenditure on naval armaments. Dr. Julio Fernandez, Argentine minister in Brazil, said Brazilians were "too silly with their megalomania and would-be imitation of the United States," while his own countrymen wished to be left in peace to build as many railroads and to make as much money as possible. He expected the arms bill would pass, and that once Argentina voted the money, Brazil might change its attitude. Fernandez also informed the British that the agreement for a mutual limitation of naval expenditure between Argentina and Chile had been prevented by Paranhos's successful diplomacy. Brazil, in reply to the propositions made in February 1908 to come into an understanding with regard to shipbuilding, indicated its readiness to consider the subject once Argentina and Chile had arrived at a definite agreement. Meanwhile, Paranhos had persuaded Chile not to come to an agreement and to preserve Brazil's friendship by not putting that country in the awkward situation of having to refuse the overtures of the other two powers. Fernandez hoped that Brazil's refusal to negotiate would change when it became evident that Argentina remained equally firm. Paranhos did not worry at all about the adverse effect on Brazil's credit in London that would be caused by his naval ambitions. The Argentine Chamber of Deputies affirmed its vote of £11 million in December 1908.[32]

The gains in naval prestige did not come without high cost. With new dreadnoughts, Brazil and Argentina in 1909 became superior to all second-class powers except Austria and Turkey. Brazil suddenly ranked ninth and Argentina tenth among the world's navies. All three countries were undoubtedly hard-pressed financially, and the naval race inhibited domestic industrial capability. Chile had the best access to European financial markets, so Santiago could borrow money in Europe more easily than either Brazil or Argentina. Chile openly stated its financial problems and hoped that a large shipbuilding program would be unnecessary. However, this did not prevent Chile from keeping the existing ships in a high state of efficiency, and spending money freely for that object. Chile and Argentina had excellent docks, but they could not accommodate a large battleship. Brazil had no docks to take the new battleships, and had not yet settled where to build one.

By now the financial strain on Brazil had begun to show. The Brazilian government contemplated the sale of its second battleship, though it did not want to dispose of the *Minas Geraes*. Brazilian officials now discussed the possibility of countermanding the order for the third dreadnought, but this would give less immediate relief to the treasury. The financial difficulties arose because revenues declined during the first ten months of 1908. Additionally, Brazil learned that Argentina planned to send its whole fleet to sea in January and to show its equivalent naval force.[33]

The Brazilian government, in need of a modernized dock facility to maintain the dreadnoughts, favored contracting with a foreign firm, perhaps from Italy or Japan. The most notable instance of domestic failure was the Brazilian-built *Tamandare*, which had proved to be a useless vessel, though it cost £2 million. Because of this, the government was inclined to abandon building a national marine arsenal itself. The Brazilians had in mind an arrangement similar to what Armstrong and Vickers had concluded with Japan, and the Brazilian government anticipated placing orders for smaller ships from yards established domestically. In 1909 Brazil accepted tenders for its docks concession. After intense competition the Vickers-Armstrong combination won the contract. Vickers bid £182,700 to deliver to Rio in eleven months a floating self-docking steel dock. Other tenders included Maryland Steel Company, £358,450;

Blohm und Voss, £307,700; Forges et Chantiers de la Mediterranée, £264,000, delivered in 20 months. Admiral Alençar voiced his support for the British proposal. However, at the last moment he granted the docks contract to the French. Armstrong officials and British Ambassador Haggard felt betrayed, and they questioned Alençar's integrity. Haggard believed that the admiral had played "an equivocal game" in allotting the heavy contract for naval docks to a French syndicate, which didn't begin any of the construction work on the docks until July 1913. Armstrong's agent consequently considered Alençar untrustworthy, and Haggard feared "that his patriotism is not entirely above being influenced by personal interest."[34]

The Brazilian government found itself attacked by its increased naval power when the navy revolted in November 1910. The crew of the dreadnought *Minas Gerais* mutinied on November 22, 1910, joined by crews from the *Sao Paulo, Bahia,* and *Deodoro.* The mutineers wanted the abolition of corporal punishment, increased pay, and less work. The ships fired shells into Rio on the morning of November 23. That afternoon *Sao Paulo* opened fire on the arsenal at Cobras Island. The government granted complete amnesty and abolished corporal punishment on November 26. In the aftermath the Brazilian navy had powerful ships, but no crews, and as a further precaution, the government removed the breechblocks from the warships so they could not fire. Although the sale of the dreadnoughts would strike a terrible blow to national pride, particularly that of Paranhos, the government began to contemplate getting rid of its dreadnoughts for a million apiece. Dr. Rodriguez, editor of *Jornal do Commercio,* Brazil's leading paper, expressed his opinion that Brazil should sell the dreadnoughts. Argentina had recently given a contract to the United States and Rodriguez feared that if the United States held Argentina to the bargain, it might be hard for Brazil to part with its ships. Rodriguez preferred to see England buy them, and he voiced these views in a meeting with the president.[35]

In spite of the strong incentive to part with its dreadnoughts, the government surprisingly augmented its fleet. Rodriguez suggested to the president that Armstrong might propose that the Brazilian government buy the ships back, giving Brazil preference for new ships to be built in substitution. The president found the idea of smaller boats, suitable to go up rivers, more appealing and a better weapon against Brazil's neighbors than the large ships. Rodriguez thought the naval minister would follow the president's and minister of Foreign Affairs's lead. He added that as dreadnoughts were useless to Brazil, this would show that Argentina's would also be useless. Moreover, the editors of *Nacion* and *Diario*—very influential Buenos Aires papers—told him that if Brazil sold its ships, they would support an identical policy for Argentina.[36] Rodriguez told Paranhos that the ships must be sold quickly, as in three to four years they would be obsolete. Paranhos assented. Haggard reported happily to London, "This is indeed a wonderful surrender on the part of the man who was answerable for the purchase and who looked upon them as the most cherished offspring of his policy."[37] However, in March 1911 the president and government decided in Cabinet not to sell the battleships. Although the president and ministers were unanimous in thinking the ships should be sold, they were afraid of the sale's effect on internal politics. Soon thereafter Chile entered the race by placing its own dreadnought orders with Armstrong. In response, Brazil signed a new contract with Armstrong for a third dreadnought at £2.675 million.[38]

Brazilian naval orders continued to flow to foreign firms. Vickers received an order for three river monitors. Early in 1912 Brazil placed an order with Italian Fiat Company for three Laurenti submarines to be built at Spezia. There was no public invitation of tenders, and competition lay between the Electric Boat Company in the United States and the Italian firm. According to rumors, the Italian success resulted from extensive bribery at the Naval Ministry. This upset the U.S. Department of State because "it is authoritatively stated that the late President promised the then US Ambassador that this order should be placed with the US and that Baron do Rio Branco [Paranhos] was incautious enough to confirm the promise in writing. Mr. Knox took this breach of faith very amiss, and instructed the US Charge d'Affaires here to remonstrate in a letter, the terms of which ruffled Baron do Rio Branco's dignity."[39]

Paranhos died in 1912. Meanwhile, the fleet had not recovered from the 1910 naval revolt. In December 1912 the British Legation received a report on the condition of the dreadnoughts from Armstrong's agent in Brazil. He reported that the Brazilians had failed to provide maintenance for their vessels. Rust and neglect had eaten away at the turrets, plates, boilers, and hoists to such an extent that Armstrong's agent estimated it would require £700,000 to put them in proper order. The third dreadnought had not yet launched by end of the year. Haggard reported, "Of all the many absurdities of the Brazilian government of late years, these ships are the most glaring instance, and the man who was responsible for this amazing and transcendent act of folly was the greatest man whom Brazil has succeeded in producing of late years, the man who when he died was publicly compared in the funeral speech, as I myself heard, with Napoleon, Alexander the Great, Bismarck, and Julius Caesar, to their detriment, the 'superman' as he was called, Baron do Rio Branco. These ships are absolutely useless to Brazil."[40]

In 1912 the death of Baron do Rio Branco altered Brazil's foreign policy. The baron had devoted all his energy to promoting the aggrandizement of Brazil. With this in mind, he had entered into a number of boundary treaties to lessen the tensions with the surrounding republics, while he sought to elevate Brazil over Argentina. He also instigated the expenditure of vast sums of money to purchase dreadnoughts. After his death, the situation changed. The new Minister of Foreign Affairs, Dr. Lauro Mueller, exhibited none of the "aggressive and dictatorial tendencies of his predecessor."[41]

By the time Brazil's third dreadnought, *Rio de Janeiro,* was ready to launch in January 1913, the navy was prepared to sell it. Russia, Italy, and Turkey had made offers in the previous six months. The Brazilian government had requested that Armstrong sell the dreadnought and, upon the sale, promised the firm another order. Brazil would accept from Armstrong payment of £2 million, the equivalent to installments already paid, but Naval Minister Alençar said he wanted Armstrong to arrange the sale of the *Rio* in concert with the British government. If Armstrong could not sell it in two months, Brazil would sell it and put up the new ship to tender. Alençar had originally supported the *Rio,* as it was first planned during his last administration. When he was dismissed following the naval revolt, his successor modified the original plan, adding to the expense. When Alençar returned to the post in 1913, he directed his first efforts toward getting rid of the ship. Before a meeting of the Chamber's financial committee, he advocated selling the ship for technical reasons. He also told the British ambassador that Brazil needed £2 million cash.[42]

Revenue from the dreadnought sale would go a long way toward easing the defense monetary burden. Brazil's authorized military expenditure for 1913 amounted to £5.635 million, or nearly 13.5 percent of Brazil's total expenditures, an increase of 6 percent over 1912. The government allocated £904,000 for purchases of war materials in 1913, up from £839,000 in 1912. Italy had been negotiating for *Rio* for some time, but another power suddenly made an offer coming through Berlin. It turned out to be Turkey offering £1.25 million cash and £940,000 paid in six weeks. In December 1913 Brazil accepted the Turkish offer.[43]

Argentine officials indicated in 1904 that they would not be influenced by external factors in their decision to maintain their military position. Speaking about the Ministry of War, Argentine president Manuel Quintana stated on October 12, 1904, "My intention is to contribute to the progress of the military institutions. Alike under the threat of war or with international peace assured, the army, its instruction and its armaments will be the dominant preoccupation of my mind." Regarding the navy, the president said, "The Argentine Republic has necessarily to be a Naval Power, by reason of its rank in South America and by the great extent of its maritime coasts."[44] Argentina's foreign minister remarked to Haggard that Brazil might be considered as a "quantite negligeable," meaning that in Argentina's eyes Brazil had no military or other importance.[45]

By 1905 Argentina still seemed unfazed by Brazil. The Argentine naval minister called attention to Brazil's plan to increase its fleet considerably in the next six to eight years, but he said the acquisition need not cause Argentina any alarm. Argentina would not neglect the defense of its own national interests and of the prominent position it held in South America. In reference to the Brazilian program, the Argentine naval minister said, "It is necessary that Congress should realize that the increase of the fleet is an insurance policy of the vital interests of the country, and it is therefore advisable not to be deterred by the expense in adopting the only means which can efficiently guarantee the free and safe progress of the Republic." The actual Argentine program was very modest. The minister proposed the acquisition of two river cruisers (800 tons), eight torpedo boats (350 tons), and some smaller vessels. The whole cost £800,000, to be paid for out of the proceeds of the sale of the *Rivadavia* and *Moreno* to Japan in December 1903. These new ships would be for internal defense and were not barred by the agreement with Chile.[46]

Brazil's action to acquire dreadnoughts caused much comment in Argentina by the end of 1905, and the Argentine naval minister in his annual report spoke of the necessity to strengthen the Argentine fleet. In view of Brazil's plans, Argentina proposed to build four large battleships: two to be ordered in 1906, and two in 1907 of 13,000 tons. Buenos Aires still had £1.512 million available from the sale of two warships to Japan. By March 1906 the Argentine naval program had expanded to include the four new battleships as well as thirty destroyers and torpedo boats.

Simultaneously Chile's program proposed to include four new battle-ships, four cruisers, sixteen destroyers, and sixteen torpedo boats. Chile formally notified Argentina of its intention to increase its fleet. Failing to come to terms with Brazil and now facing Chile's naval program as well, the Argentine government sent Rear Admiral Domecq Garcia to Europe, and Captains Manuel Barraza and Juan A. Martin prepared plans and es-timates of the new ships.[47]

Perceiving the blossoming opportunity in the South American naval market, various foreign firms prepared to enter the fray. In an ironic twist, this time the Germans complained about the "unscrupulous" methods of the British firms, who allegedly lavished very generous gifts on officials to keep the market for themselves. Dr. E. Cerio of Capri, a naval engineer and technical delegate from Krupp, arrived in Buenos Aires from Chile. He was joined by Admiral Galster, a retired German naval officer also employed by Krupp. They both had offices at Tornquist's, the major financial house that served as Krupp's financial and permanent representative in Argentina. Both the German and Italian governments began trying to exert their in-fluence in favor of their respective shipbuilders. Leopold Perez, Vickers' agent, was the only representative of a British shipbuilding firm in Ar-gentina. According to Cerio, neither Chile nor Argentina wanted to take the initiative in ordering new warships in defiance of the existing disar-mament convention. Cerio attributed Krupp's failure to get an order from Brazil for the new armored ships to the financial power of Rothschild in London, who had made it a condition in making the loan that the ships be built by English firms. In Argentina and Chile the Germans found it very difficult to gain warship orders because German systems would disrupt the unity of the fleet, especially for guns and ammunition. Dr. Estanislao Zeballos, the minister of foreign affairs, practically assured Townley, British ambassador to Argentina, that the large battleships would be or-dered in England although orders for some smaller ships would have to be placed in Italy. Argentina had also received an offer from Japanese ship-builders asserting that they were able to build warships on equal terms with Britain or any other European power.[48]

Tornquist led the opposition against the naval race with Brazil. As dis-cussed above, he had attempted to negotiate with Brazil behind the scenes. Tornquist also labored very hard in Congress to stem the tide. He managed to convince a number of his fellow members on the Armament Commission, who had been in favor of the purchase, to oppose it. Ex-president Roca also opposed the purchase. Originally, Tornquist had been the sole opponent, but by December 1906 the majority was moving his way.[49] Roca had no fear of Brazil. As he told Haggard, "They might build as many ships as they liked; it pleased them and did not hurt the Argentine."[50] In January 1907 the Argentine Congress ended its session without passing the bill authorizing the government to increase naval strength and to construct fortifications. General Roca and Tornquist had won a political victory, at least in the short run.[51]

As a deputy in the Congress, Tornquist's opposition to the naval bill and purchases of new artillery batteries put him at odds with his business interests as Krupp's representative in Argentina. By April 1907 the German minister was lobbying very hard that the battleship orders be given to Germany. The British unfairly believed that Tornquist had strenuously prevented the vote in Congress because the order for battleships would probably have gone to British firms. A delay would allow the German firm to play for time and possibly gain an endorsement from General Roca. Roca had ardently maintained that no new ships were needed, but the German minister in Argentina showered his attentions on the general, and Roca was feted in Berlin.[52]

Tornquist had become a liability to Germany. By March 1908 Julius von Waldthausen, the German ambassador in Argentina, reported to Berlin that the Argentine minister of foreign affairs and the war minister had told him that the president and his government were extremely disgruntled with Tornquist's opposition to the current government. While undoubtedly Tornquist had performed great service for German interests in the past, more recently in the Chamber he had become more enmeshed in Argentine internal politics. In conflicts between Congress and the Cabinet, Tornquist often sided with the Congress. Tornquist's vocal opposition to the president, the minister of the interior, the war minister, and especially Foreign Minister Zeballos impressed on Waldthausen that the current government would not grant any more orders to Krupp so long as Tornquist represented that firm.[53]

Political support from the Argentine government for the naval expansion continued to build. Zeballos told Townley the matter would probably occupy the early attention of Congress. The president and his ministers favored acquiring the battleships, not to counter Brazil, but rather because Argentina had need of a stronger navy to maintain its present position in the world.[54] The congressional vote empowered the government to buy two battleships, six torpedo boat destroyers, and twelve smaller vessels. If necessary, the government could also buy a third battleship and three coastal destroyers. Of the £11 million approved, naval armaments claimed £6.4 million and the army received £4.448 million.

Even with a reasonable expectation that the battleships would be ordered in England, the British embassy worried about the possibility of financial instability in the country and considered the program a reckless expenditure on armaments.[55] As Townley lamented to London:

> If it is finally decided to spend large sums on naval and military armaments, it is obvious that there will be little or no money available for the various excellent schemes now awaiting the sanction of Congress, which are intended to develop the National Territories and other rich districts of the country, by means of light railways, roads, and irrigation works, whilst there will be no question of reducing the present high customs duties which contribute so much to make the rate of living an almost impossible one for the laboring classes.[56]

Argentina had a balanced budget at the time; therefore the government did not need a loan to pay for the battleships. Naval expenditure budgeted for 1909 amounted to £1,448,168, and for 1910 the sum had increased to £1,543,470. Additionally, Congress had appropriated a special grant for fleet renovation of £1.572 million. To pay for armaments expenditure of £1.31 million in 1909 and double that in 1910, Argentina's finance minister noted that surplus revenue would suffice. The first installments of payment were already deposited and were drawn from general revenues.[57]

Initially President Figuera Alcorta had wanted to invite only British shipbuilding firms to tender. Townley met with a member of the Technical Commission, who said the condition of vessels now forming the Argentine navy had been steadily deteriorating since 1904 because of insufficient funds for the upkeep. The Argentine naval officer added that the ships built in Italy had proved most unsatisfactory from every point of view, and therefore a unanimous opinion from the commission for building in England could be expected. At first only British firms received invitations. Later, Zeballos convinced the president to expand the field to include German and Italian factories. With Alcorta's assent in the latter half of July 1908 the government approached Vickers, Armstrong, Germania, Vulcan, and Ansaldo about preliminary prices.[58]

In 1909 the Argentine Naval Commission under Admiral Domecq Garcia accepted tenders from European and U.S. firms for torpedo boat destroyers and battleships. Settling the destroyers first, to appease all parties, Argentina divided its contracts equally among English, French, and German firms, each receiving four destroyers. To preserve unity of armament, the destroyer contracts excluded guns and ammunition. The armaments would come from the firm that won the battleship orders. The contest over the battleships was even more heated, and plans were accepted, altered, sent back to London, and generally so transformed that the vessels grew from 20,000-ton ships to vessels of 27,000 tons when fully loaded. In the end U.S. firms secured the contracts. The Fore River Shipbuilding Company (Quincy, Massachusetts) received the order for both ships, in conjunction with the New York Shipbuilding Company (Camden, New Jersey). The Americans had consistently kept the price below that of all other competitors, and the U.S. ambassador lobbied hard, paying calls almost daily. The final test resolved itself into an application from the Naval Commission for plans and specifications from six selected firms showing the time within which they were prepared to construct a vessel of 27,500 tons at a cost of £2.2 million. Fore River Shipbuilding proposed to build a ship in the shortest amount of time for the lowest cost. Armstrong proposed a longer time and higher price, but the British firm had a powerful following among the naval experts. After a prolonged and heated debate, the U.S. firms earned a tie vote of 6–6 and the naval minister as chairman cast the deciding vote in their favor. Townley complained back to London:

> There is good reason to believe that the American Minister resorted to every sort of political pressure, and that a promise was given that if the ships were built in the United States Brazil should give no trouble until they have been delivered. It is hard to say what Argentina will do with their big ships when she gets them, as she has no port into which she can take them, and even were the money forthcoming for the construction of a new port, or for the enlargement of an old one, there would hardly be time for such works to be completed before the ships are delivered, if the two years limit is strictly adhered to. British naval contractors say that it is quite impossible that the ships can be ready in the time.[59]

The stunning American success in landing the battleship contracts owed much to the efforts of the State Department. No U.S. firm had ever built a warship for a South American country before, whereas in the previous twenty years British firms had constructed thirteen cruisers (distributed to Argentina, Chile, and Brazil), the Italians had supplied four cruisers (Argentina), and the French one cruiser (Brazil). Moreover, no staunch proponent for the United States could be found among the key decision-making figures. The Naval Commission was generally pro-British, the vice president looked to Italy, the minister of war advocated for Germany, and the minister of marine was decidedly anti-Yankee. Understanding the difficulties involved, Philander Chase Knox, President William Taft's secretary of state, took the initiative by arranging for U.S. bankers to subscribe to a $10 million loan in anticipation of financing the contracts. The newspaper *La Prensa* helped turn the situation in favor of the U.S. firms by questioning the integrity of the Naval Commission. The paper reported the attempt by some commission members to obtain gratuities from a British firm for torpedo craft. Under public scrutiny the Naval Commission subsequently reordered the rankings of the tenders, with the Americans now placing first, displacing the Italians from first to sixth. The British remained second in both rankings. British pressure on the Naval Commission then led to a call for new bids for a ship with specifications corresponding to English bidders. Although the Americans submitted the lowest price, the British managed to get the Naval Commission to allow Armstrong-Vickers to lower their price by $570,000, underbidding the Americans. Only the State Department's diplomatic pressure on the Argentine president forced another competition, which the Americans won. On January 21, 1910, U.S. firms signed the contracts for two battleships with a short-term option on a third dreadnought.[60]

British skepticism about the U.S. delivery schedule for the dreadnoughts turned out to be correct. The *Rividavia,* under construction at Fore River Shipbuilding Company, should have been ready for delivery and sea service on February 5, 1913. The guns held up the work. As of March Bethlehem Steel had not yet delivered the twelve-inch guns, and as a result the turrets required another five months. The sister ship *Moreno,* building at New York Shipbuilding Company, was even further behind. Fewer guns had been delivered at Camden.[61]

The destroyer program also experienced difficulties. The Argentine government refused to accept the four British destroyers built by Laird on the grounds that the vessels did not meet the technical requirements. Laird subsequently sold the destroyers to Greece. To replace these British warships, Argentina turned to German suppliers. The Naval Commission evaluated tenders from Germania and Schichau. Germania got the contract, and Bethlehem Steel provided the guns and ammunition. Meanwhile, the French drew public criticism in *Nacion* for their tardiness and inferior performance. The French firm Brosse et Fouché, the contractors for the destroyers *Mendoza, Salta, San Juan,* and *Rioja,* had not met the contractual requirements for speed or coal consumption. Captain Lagos, head of the Naval Commission, journeyed to France to make arrangements with the firm and "found it impossible to arrive at anything practical while defending the legitimate interests of his country."[62]

With Brazil's acquisition of *Rio* and Chile's newly ordered ships, Saenz Valiente, the Argentine minister of marine, announced in June 1912 that the moment had come to procure the third dreadnought. Twelve senators drew up a memo, addressed to the chief executive, recommending immediate purchase of a third dreadnought. In debate, the ministers of foreign affairs, navy, and finance all spoke in favor of the memo, which gained approval, 15–7. However, after General Roca returned from Rio the president decided to postpone consideration of the matter.[63] The Brazilian threat evaporated when Brazil halted work on the *Rio,* and then sold it to Turkey in December 1913. Argentina reacted by not acquiring a third dreadnought, and the second South American naval race ended.[64]

Meanwhile, in Chile increasing concerns about Peru led the admiralty of the late President Riesco to contemplate a considerable increase of the navy. In August 1906 General Salvador Vergara, naval minister, outlined a naval building program that consisted of four battleships and three fast cruisers over ten years. Vergara assured the British embassy that the tenders from British firms would receive fair consideration, but he added that Admiral Joaquin Munoz, a member of the Chilean Naval Commission in Europe who had recently been in Germany, had been very much impressed by what he had seen in the shipbuilding yards in that country. The government had received tenders from five British firms, two French firms, one German firm, and one U.S. firm. Currently the government hoped to order two battleships and one cruiser. Armstrong, Vickers, Brown, and Fairfield remained under consideration. Chile's new president, Pedro Montt, disapproved of the expenditure on naval construction, considering that such funds would be better spent on improving internal communications.[65]

In August 1907 Congress voted to expend up to £5 million on new naval units. Regarding the new armaments, the government requested freedom to contract with any foreign firms that offered the best terms. Admiral Jorge Montt, director general of the fleet, worried that the whole program involved more money than anticipated. Although naval officers were anxious for the proposed fleet augmentation, he hesitated to recommend a program for which the cost would be opposed by the president and in political circles.[66]

Admiral Montt personally preferred English-built vessels and materials, as he had ordered on his own responsibility new boilers from England (Babcock and Wilcox type) and countermanded the French Belleville boilers to refit the *Prat*. British ambassador Ernest Rennie believed that the "Admiral's prejudice in favor of British built vessels is generally shared in the expert circles of the Chilean navy, and that, were the decision allowed to rest with Naval men alone, there is little doubt that contracts for naval construction would go to British shipbuilding yards. In the case, however, of larger contracts, much political and outside pressure in favor of continental builders is brought to bear, as was seen last March, when the more extensive program was under consideration."[67]

Chile edged into the dreadnought race gradually. The collapse of the nitrate market in 1907 and the earthquake in 1908 created poor economic conditions in the country. In response to Peru's addition of two new cruisers and dock at El Callao for construction of torpedoes, Chile considered an increase in its navy by two new cruisers and twenty-five torpedo boats. President Montt now wanted to buy two battleships and a cruiser in England with a loan of £3 million. In 1908 the government first broached the question of the need of one dreadnought, but congressional discussions were negative. The government looked to add one cruiser and repair existing ships in the ensuing year. The dockyards at Talcahuano could now repair ships up to 20,000 tons, thanks to the floating dock supplied by Germany. In 1910 the Chilean Chamber of Deputies voted that the president could enter into a loan for £4.480 million over five years and raise revenue £400,000 yearly to purchase one dreadnought (22,500 tons), three destroyers, and two submarines. Additionally, Congress authorized £1 million for coast defenses and £80,000 on naval arsenals. Within eighteen months Chile intended to purchase another dreadnought and three more destroyers. In October the government gave the French firm Allard Dollfus Sillard and Winot a contract to construct the Talcahuano naval dock for £948,000.

Germany had offered to lend Chile the money needed for ship construction on the condition that German firms received preference. Vickers, in response, lined up Lord Rothschild to finance Chile if required.[68]

Ultimately, Britain retained its dominant position in the Chilean naval market. In 1911 British firms claimed all naval orders with the exception of two U.S. submarines (Holland Company). The contract for six destroyers went to the English firm J. S. White of Cowes. The competition had been fierce among British, German, and U.S. firms. However, the Chilean Admiralty stacked the deck by stipulating that all armament must be British. When the German minister argued for the superiority of German ships, Admiral Montt replied, "We believe British-built ships to be the best in the world. When you beat them, we shall be quite ready to buy our ships in Germany."[69]

Chile decided to order a second dreadnought in 1912. The United States made every effort to secure this battleship, but Armstrong landed it in May 1912. Once again the determination of the director general of the Chilean navy, Admiral Montt, to have no dealings except with British firms meant that the contract was a foregone conclusion. The tenders for coastal artillery included Armstrong, Vickers, and Coventry Ordinance Works. Bethlehem Steel underbid everyone, and Chile thought it expedient to give the order to Bethlehem as a matter of policy, since the Americans were "very sore about having failed to secure second Dreadnought, in spite of very advantageous terms offered. Montt openly expressed to HM Legation his regret that this course appeared in the circumstances to be the wisest, since he would have preferred to have nothing but British material, but the expediency of not giving offence to the US had also to be taken into account."[70] Santiago now had building in England two dreadnoughts and six destroyers at a cost of £7 million.[71]

The primary mission of the Greek navy was to safeguard against a raid by the Turkish fleet. Given the dilapidated state of the Ottoman navy in 1904, the Turks did not pose a serious danger. The Turkish fleet consisted of only one reconstructed battleship, two second-class cruisers, and two destroyers. The suggested Greek naval program in 1904 called for building three battleships and eighteen destroyers. Confronted with such a Greek force, the only hope the Turks could have of inflicting any damage on the Greeks would depend on their fast cruisers. Very likely the lone Turkish battleship would not last long among the islands of the archipelago if the Greek flotilla was in existence. To win in the scenario the Greeks required cruisers faster and stronger than their Turkish foes, since "the three proposed battleships would be laughed at by the Turkish fast cruisers, who would be free to work destruction at their will."[72]

For the new naval construction the Greeks had 10 million drachmas in the naval budget, with 3 million drachmas coming in 1904. The principal source of funds for the Greek navy derived from the interest of the Naval Defense Fund. This privately collected fund had as its core a large legacy left to the country for the purpose of improving the navy. Annually the Greek Chamber added to the fund by a vote of 5 million francs (about £200,000). These funds were allowed to be accumulated or expended as appropriate.[73]

British firms expressed interest in Greek construction. A company representative from Fairfield Shipbuilding met with Prince George to see about naval contracts. The prince informed Fairfield that the Greeks wanted three battleships, costing £450,000 each, and fifteen destroyers. They preferred to place the order with a single firm, with payments spread over ten years. Laird Company had already offered these terms but wanted 20 percent paid in advance and 5 percent interest. Since Prince George wanted to avoid any loans, he had ruled out Laird. Fairfield's representative did not like the ten-year spread, and the Greeks did not make an offer to Fairfield.[74]

Greek naval acquisitions proceeded slowly. During 1905, the naval program was reduced to twelve destroyers. Proposals for the three battleships had made no progress because the contractors consulted disliked the terms for complete payment. While negotiations stalled on this account, the government postponed its decision on battleships. The government did send a committee to France, Germany, and England to decide where to order destroyers, and the Greeks invited tenders from two Italian firms. In October 1905 Greece ordered two destroyers (each 300–350 tons), one from Vulcan (1.2 million francs) and one from Yarrow (1.35 million francs). The Greeks were also talking to Stabilimento Technico in Trieste about a destroyer.[75] According to War Minister Demetre Rhallys, the Yarrow boat cost more than the Vulcan boat, even though the German firm also included torpedo tubes and fittings in their price and the English firm did not. Alfred Yarrow suggested to the Greek government that it should set up the necessary appliances to build destroyers for itself, but the Greeks realized that it would be "much cheaper to buy them."[76] By September 1906 the Greek government had given further orders abroad for new destroyers. After these vessels were completed the Greek navy increased by eight destroyers, of which four were British (Yarrow) and four were German (Vulcan). None had yet arrived in Greek waters.[77]

During 1907, Greek ideas regarding its naval program had undergone considerable development. Early in the summer retired French admiral François Fournier visited Athens. Fournier, an enthusiast for small craft, and especially submarines, had conversations with the king and impressed the monarch with his views. The king obtained sanction from France to have Fournier reorganize the Greek navy. Fournier's plan for the fleet involved destroyers, torpedo boats, and submarines, together with three old cruisers. Greek naval opinion opposed the French admiral's plans.[78] In particular, the prime minister and Greek naval officers categorically rejected Fournier's plan calling for £2 million for submarines, and their opposition forced Fournier to abandon the idea. The British ambassador reported to London, "The Crown Prince also told me that the scheme was hanging fire and that Monsieur Clemenceau was correspondingly indignant."[79] In December 1908 the Greek prime minister confided to the British that frustration of the appointment of Admiral Fournier had entailed the loss of the Greek loan from French sources, and as a result the country's naval preparations were also stalled. "This statement of the Prime Minister is a clear indication of the closeness of the connection in the eyes of the French Government between the commercial and the political aspects of their relations towards their Greek clientele."[80]

The French diplomatic pressure did not translate into naval orders, as German, British, and Italian firms won contracts over the French. Greece began major naval expansion in late 1909. In November the Greek Chamber ratified the bill for a contract to purchase from Orlando and Company of Livorno an armored cruiser of 10,120 tons displacement for 23,768,844 francs. The *Averoff* would be delivered no later than June 1910. Greece paid for the cruiser with 7 million francs of the Averoff legacy and borrowed the rest. The Greek government signed the contract in December, and Armstrong supplied the armament. Yarrow competed for destroyers and offered the Greeks two destroyers for delivery in eight months. However, Vulcan proposed two similar vessels in only three months' time by selling to Greece the two destroyers already being built for the German government. Vulcan won the contract.[81]

During the later years of Abdulhamid's reign, the German hegemony in Ottoman naval orders had already faded. In the reconstruction of 1903–1907, Ansaldo of Italy modernized three ironclads. Only one iron-clad contract was handed to a German firm in 1903–1907 (Krupp), and this was only because Ansaldo had been unable to complete the work in 1899. Other European suppliers moved into the German domain of tor-pedo boats.[82] In January 1905 the Turks bought four gunboats from Schneider. Later that fall Schneider received an order for four torpedo boats.[83] The following year Schneider gained another contract for nine coast patrol boats valued at TL223,500 and one gunboat valued at TL71,500. Ottoman naval orders amounted to 13.8 million francs.[84] These purchases expended the money retained by the Ottoman Bank for orders given to French firms, according to the conditions of the loan concluded in the spring of 1905 with the imperial Ottoman government. The patrol boats were designated for service on the Red Sea. By 1908, the Ottoman fleet possessed fifteen torpedo boats, including seven from An-saldo (1904–1906), four from Schneider (1906), and two from Ansaldo (1900–1901). The Turks also ordered four U.S. submarines, Holland type, in 1908 at TL50,000 each.[85]

Armstrong hoped to make a major sale to the sultan, and had been try-ing since 1904. J. Falkner visited Constantinople on behalf of Arm-strong, Whitworth and Company in June 1904. At that time the sultan allowed British ambassador Sir Nicholas O'Conor to present Falkner personally. Abdulhamid said he would give Armstrong an order for a warship, and the sultan requested that Falkner send the ship's plans di-rectly to him. Falkner delivered the designs without any result. Subse-quently O'Conor reminded the sultan of his promise, and was again assured that the matter had not been forgotten. Again in October 1905 O'Conor brought up the subject, and Abdulhamid issued a decree in-structing the admiralty to order from Armstrong vessels in value of £7.300 million or more, according to the balance available in the budget for the next fiscal year. When the loan negotiations were in process in 1905, O'Conor had considered making Armstrong's case when orders were being given to Creusot as one of the conditions of the last Turkish loan in Paris. The British ambassador remained silent, though, since he "was unwilling to take any action which might be interpreted as an un-friendly act by the French Embassy, especially without feeling certain that it would serve our interests."[86]

The British embassy disliked the idea of lobbying the sultan for British naval orders. In 1906 O'Conor expressed reluctance to urge Abdulhamid "to buy ships which the Treasury cannot afford, in its present impoverished state, without serious injury to the country and which are destined like the former cruisers ordered from Messrs Armstrong and from Messrs Cramp of the United States, to be anchored in the Golden Horn till the machinery becomes useless, the crew incapable, and the ships worthless as an engine of war. I do not think it is consonant either with our national character or policy to press such requisitions unduly upon the Sultan or his Government."[87] The following year O'Conor still opposed pressing the Ottoman government or the sultan to order a cruiser from Armstrong, "when the issue of such orders cannot be regarded otherwise than as a flagrant abuse of diplomatic influence in view of the impoverished condition of the Imperial Ex-Chequer, the urgent straits in which the Administration finds itself and the known fact that the three last cruisers acquired, one of which was ordered from Messrs. Armstrong, have been deteriorating in the Golden Horn for the last three years."[88]

The sultan surprised the British in the summer of 1907 when the Turkish government began discussions to buy a cruiser from Ansaldo of Genoa. On August 31, 1907, Mario Perrone of Ansaldo and Ottoman naval minister Hassan Rami Pasha signed a contract for the construction of the cruiser *Drama* for £330,000. The dealings involved several pecuniary improprieties equivalent to 10–15 percent of the total value of the contract, including a commission of £20,000 to Rami and Izzet Pasha, and an additional 10,000 for Halil Pasha.[89] O'Conor sent a message to the sultan reminding him of his promise to Falkner. O'Conor then learned that U.S. ambassador John Leishman also had received a promise from the sultan regarding the eventual construction of a cruiser in the United States. Leishman, who recognized the priority of Abdulhamid's promise to Armstrong, went to the palace to try to block the negotiations with Ansaldo. As a means to bolster his representations Leishman warned the sultan of the possibility that Armstrong would claim damages. Leishman's actions put a temporary stop to negotiations with Ansaldo. However, in August the Ottoman government informed O'Conor of its decision to purchase a cruiser from Ansaldo. At this stage the British embassy took no further steps regarding Armstrong's order. Acting on the assessment of the British Admiralty, which reported that Armstrong had amalgamated with Ansaldo in 1904 and the British firm held one-third of the joint capital, the Foreign Office believed that Armstrong could safeguard its own interests through Ansaldo.[90]

Again, British diplomats proved reluctant to apply pressure on behalf of an English firm. Armstrong still wanted the embassy to press the sultan for the order. The firm pointed out that Lord Armstrong played no part in the management of Ansaldo. Although Armstrong had good relations with the Italian firm, the British company exercised no control over its management. Moreover, Armstrong's holdings were not large enough to allow it to give an order to the Italians. From Armstrong's perspective the Turks had systematically ignored English firms for many years while other foreign manufacturers gained important contracts.[91] By the end of the year, the Foreign Office informed O'Conor that since the Ansaldo order would probably not materialize, "His Majesty's Government have no wish to embarrass the Sublime Porte by pressing for an order for Messr. Armstrong at an inconvenient time."[92]

The Young Turk revolution brought a change in naval affairs. By deposing Abdulhamid II in 1909, the Young Turks had removed the single most important impediment to revitalizing the Turkish navy. Accordingly they made plans for a major improvement in the composition of the fleet. The new program called for six battleships, twelve destroyers, twelve torpedo boats, and six submarines.[93] In September 1908 the new Young Turk government requested from London an English admiral to oversee reorganization of the Turkish fleet. The admiral would have absolute power to design for naval artillery as well as torpedoes. The Foreign Office considered the loan of a British admiral to Turkey as "one of the highest political importance," according to Edward Grey. The British government sanctioned the appointment of Rear Admiral Douglas Gamble to the Turkish navy in October. The British wanted Gamble to have direct access to the minister of marine and authority under him to deal directly with all matters relating to the navy and dockyard to carry out reorganization.[94]

As a mark of Britain's improved position in the Ottoman market, the Turks employed Admiral Henry Woods of the Royal Navy to advise them on naval modernization. Woods's report on the Turkish navy in October 1908 noted its rundown state and advocated new construction as imperative. Significantly, Woods did not take as his job maximizing business for British shipbuilders. He reported, "This program must not be too ambitious. It should be drawn up with regard to the amount that can be set aside from the revenues of the country as grants for construction . . . Too extensive a program would be a mistake, on account of the onerous burden that would be cast upon the revenues." The British admiral recommended the construction of two armored cruisers, twelve destroyers, thirty torpedo boats, and eight submarines.[95]

Turkish respect for British shipbuilders had grown during the first decade of the twentieth century. In November 1908 Captain Williamson reported to the British ambassador Sir Gerard Lowther that all the Turkish officers

> were fully determined to come to England for their new ships. They are thoroughly dissatisfied with the American-built *Abdul Mecid* and recognize that the huge sums of money spent on the reconstruction of the old ships might almost as well have been thrown into the sea, the only people gaining profit from the work being Ansaldo and the other firms which took it in hand, together with those in the Ministry who arranged the contracts. On the other hand, the Armstrong-built *Abdul Hamid* has given great satisfaction, and the wonderful good work put into the hulks of the old iron-clads by the Thames Ironworks and other English firms has quite impressed upon the Turkish officers the great superiority of English work over that of an other nation. They turn up the whites of their eyes in horror at the large amount of bribery and corruption which went on under the old regime.[96]

Since the Turkish government lacked money, the grand vizier thought it unadvisable to engage Gamble's services before being able to fund naval reorganization.[97] Despite the vizier's reluctance, the naval minister eagerly wanted to proceed with British assistance. The Turkish Admiralty had drawn up a scheme of construction that entailed an expenditure of some £18 million and included the building of some six dreadnoughts. This estimate would be submitted to Parliament, which had to devise some way to find the money, but the naval minister assured Lowther "that every penny of it was to be spent in England, not a rivet even coming from Germany or other countries, with which they had had ruinous experiences during the late regime."[98] The Turks wanted to follow the Japanese example and have their navy reorganized purely on English lines. In December 1908 the Turks accepted Admiral Gamble with an annual salary of £3,000.[99]

British officers and diplomats continued to discourage major Turkish naval purchases. In April 1909 Admiral Gamble wanted only a modest naval credit for six years to build one small cruiser, torpedo catchers, torpedo boats, submarines, and a minelayer. After the question of approval of the extraordinary credit would come the question of which firms would compete. By July Turkish naval minister Aarif Hikmet Pasha reported that the credits for six submarines, Holland type, had been granted. The choice of Holland type originated with Gamble.[100]

Even as the Turks nervously watched Greece acquire the *Averoff*, the British tried to forestall a Turkish naval race with Greece. A report to Edward Grey stated:

> The Porte are in a state of panic with regard to the acquisition of the new cruiser by Greece; for it will mean their certain defeat at sea: indeed it would be out of the question for them to join issue with the Greek fleet. The Porte are convinced that the military party mean war: they are pressing for the delivery of their new cruiser in July. Admiral Gamble fears that the Porte may entirely lose their heads and purchase the Brazilian battleship—a super dreadnought—now building here for 2 million pounds. It would be useless to them as they have no one capable of commanding her and no engineers who could understand her machinery. He is doing his best to dissuade them.[101]

At this time the Germans also discouraged the Turks. In mid-December 1909 the Turks considered the purchase of a German warship. The Turkish ambassador in Berlin, Osman Nisami Pasha, made inquiries regarding a 13,000-ton battleship and 350-ton torpedo boats. The Greek acquisition of the *Averoff* and its effect on the Cretan question motivated the Turks to find a countermeasure. The German military attaché in Constantinople, Major Walter von Strempel, disapproved of the idea on the grounds that the Turks would not be able to handle such a new ship, and that English officers would gain access to German naval designs.[102]

Turkish preoccupation with Greek naval power drove policy. According to Grand Vizier Hakki Pasha, Turkey had to increase its navy to cope with Greece. He told Lowther, "No objection could be made to this, as it would merely be a defensive measure and act as a restraining influence on Greece, it being obvious that the [Great] Powers would never allow Turkey to make an unprovoked attack on Greece."[103] In mid-January a new naval minister, Halil Pasha, took office. Halil was an experienced naval officer who had also served as naval attaché in London. Although he had pro-British sympathies, Halil clashed with Gamble over decision-making control regarding new construction. As a way to assert his authority, Halil made Gamble step down. Halil then entered into secret and unauthorized negotiations with Armstrong for the purchase of two dreadnoughts and a pocket battleship as the means to neutralize the Greeks. In April he announced a competition for naval orders in which

German firms were invited to tender. The Germans, aware of Halil's discussions with Armstrong, regarded the competition as a farce. Meanwhile, the pro-German war minister, Mahmud Shevket Pasha, feared that the failure of the Germans to supply warships would be regarded as a personal defeat for him. He urged the Germans to sell their warship *Blücher* without placing any stipulations for further orders in Germany. Otherwise, the war minister warned, Germany would be closed out of the Turkish naval market for years to come. When Halil came forward with his plans to purchase the warships from Armstrong for TL5 million, the grand vizier forced Halil to resign on May 29, and he halted the order on the grounds that such a purchase far exceeded the state's financial resources. More likely, the grand vizier found the unavoidable wait for the construction of a new dreadnought intolerable, and he wanted to procure an existing German dreadnought immediately so that Greece could not counter with anything comparable. German kaiser Wilhelm II opposed such a sale as it would alienate Greece and bring on a clash with England in the Mediterranean.[104]

During the summer of 1910 the Turks closed in on their goal. On July 13 Shevket Pasha informed Admiral Williams that the Turkish cabinet wanted to purchase a ship capable of dealing with the Greek warship *Averoff* as soon as possible, and he "regretted that we had some months ago declined to sell them a ship and asked whether we had not changed our minds."[105] The British had not changed their minds, but by June the Germans had started to reconsider the merits in selling warships to Turkey. If Germany continued to oppose the Turkish request, the Germans ran the risk that Turkey would return to its initial plan and place orders exclusively with Britain. In turn, this would offer assistance to England in building a rapprochement with the Turks as the English could blame the Germans for opposing Turkey. However, a naval sale to Turkey could facilitate drawing the Turks closer to Germany and Austria-Hungary, as the kaiser desired. For these reasons the German Foreign Office urged the navy to accede to the Turkish wish.[106] Sensing that the Germans were growing receptive, Shevket Pasha lobbied them more forcefully in July. The Turkish war minister explained the opposition he faced in the cabinet from ministers who had come out very much for England. He laid out a scenario wherein if Germany did not make a move, the German shipbuilding "would be driven from our market for ten years . . . The agreed 5 million pounds must go to England if an order for ships is handed out. Only the sale of a ready ship would now make possible a deviation from that."[107] In mid-July the Germans proposed the possible transfer of the two old Brandenburg class armored cruisers. The grand vizier was well disposed to this solution since his main concern was inhibiting the Greeks from opting for war. On August 5 the Turks bought these ships.[108]

Two important events in 1910 helped bolster the German profile in Turkey. The first was the sale of the two warships. Second, while France refused to sanction a loan, and Britain abstained, Germany came through with the loan. The German loan amounted to TL11 million at 4 percent interest. Relying on the regular budget proved inadequate since for 1910–1911 expenditures exceeded revenues by more than TL7 million (TL32 million/25 million).[109]

In diplomatic circles the Turks played down their naval acquisitions and sought to avoid the appearance of aggressive naval expansion. While in Berlin the Turkish finance minister Cavid Bey told British ambassador George Goschen that the Turks had intended to order five gunboats in France for a long time, and he anticipated the order would be given soon. Cavid discounted rumors that the Turks were considering buying any more ships from Germany or England. The two German ships Turkey had bought "were necessary to take down the pride and confidence of the Greeks in their navy, a pride and confidence which, if allowed to run riot, would have done harm."[110] In the Turkish press it was a different story. The newspaper *Ikdam* elaborated on how the recent acquisition of two German armored cruisers would affect the balance of power in the Mediterranean. "Before the acquisition of the two cruisers, Turkey's Navy was without importance; that of Greece, by the addition of the *Averoff*, was superior to ours. If the purchase of a couple of old German ships is enough to give us this position, what will be the extent of our influence when we buy new ships and build ships ourselves? All the [Great] Powers will change their policy towards us."[111] As it happened, the public opinion proved closer to the truth.

Two English firms, Palmers and Armstrong, competed to build the dreadnoughts ordered by the Ottoman government. Palmers Shipbuilding had first sent its agent, Gerald F. Talbot, to Constantinople in December 1908 seeking the contracts.[112] By June 1910 the U.S. firm Bethlehem Steel had allied with Palmers to tender for the Turkish battleships. Bethlehem needed an English partner because the Ottoman government had adhered strictly to its policy of building these ships in England, notwithstanding the protests of German, French, Italian, and American builders.

The Palmers-Bethlehem collaboration had given the best proposal based on price, and the Turkish Commission had accepted it and passed it on to the Council of Ministers for their sanction. Bethlehem received notification of the council's acceptance on June 26 subject to Parliament's ratification. The Turks indicated that there was no question about this. Oscar Strauss, the U.S. ambassador in Constantinople, had been trying to see Grand Vizier Hakki for three weeks, but had been unable to do so. Thus, Bethlehem had had no assistance whatever from the embassy. The matter would go before the Council of Ministers the following week. As J. E. Mathews of Bethlehem wrote to Leishman, the U.S. ambassador in Rome, "the opposition from England will be very strong and unless it is pushed through at once I fear it will never get through."[113]

The U.S. ambassador cautioned against overly optimistic expectations about gaining the Turkish contracts quickly or easily. Leishman anticipated that the Vickers-Armstrong group with its banking support would have the advantage since the Turks, as usual, were strapped for cash. "Anyone not thoroughly acquainted with the Turks is apt to be deceived as they are always on the point of settling things the next day, but they generally drag along for months—sometimes, years; and occasionally, into eternity—and low prices for perspective business of this kind is not attractive."[114]

Leishman's assessment proved prescient when the Turks communicated their interest in buying new battleships to Armstrong's agent J. M. Falkner in Constantinople in February 1911. By March 2 the Turkish naval minister had signed the preliminary agreement for an order of two battleships to be placed with the Armstrong group.[115] The naval minister in his statement to the Chamber of Deputies on March 25, 1911, indicated that he was disposed to order hulls from Armstrong if they reduced their price TL150,000, which was the difference between Armstrong's price and Palmers'.[116] When the offer by Armstrong received preference, Palmers at first lodged a complaint with the ministry and afterward took it to the chamber. Talbot attributed his firm's loss of the contract to the negative influence exercised by Lowther. Although the Foreign Office had issued a letter March 6 in which the British government stated that it had no preference between Armstrong and Palmers, Talbot heard from a variety of Turkish ministers and representatives that British ambassador Lowther had made confidential and verbal communications to the Turkish government undercutting that position. Talbot had direct statements made by the Turkish ministers of finance and marine in the presence of witnesses, and similar statements made by Talaat Bey, head of the Union and Progress Party in Parliament, that Lowther had indicated that the British Foreign Office had sent that letter only under pressure from Palmers' directors, and that it was the desire of the British government that Armstrong should build the ships. "We have been defeated purely and simply by our own Ambassador's partisanship for Armstrong, and by his grossly unfair treatment of ourselves," wrote Talbot in a letter forwarded to Grey.[117]

Doubts about Palmers existed on the British and U.S. sides. No evidence ever came to light that Lowther had acted improperly or with hostility toward Palmers. Although the British Admiralty and Foreign Office did not consider Talbot's complaints "reliable or credible,"[118] the British government did have an objection to Palmers. Unlike the Armstrong group, which consisted entirely of British firms, Palmers' proposal had been to build battleships in England but to supply them with armor and guns from Bethlehem Steel. The firm was working with an American agent and the support of the U.S. Embassy.[119] As a British Admiralty memo noted, "I am reluctant to commit us more than is absolutely necessary to support Messrs. Palmer, because whenever we do so we shall have to reckon with the fact that they have gone to America for their guns and plates, so that the same kind of difficulty may arise."[120] From the American side, Leishman, too, had doubts about Palmers. "I haven't much confidence in the ability of Palmers to do much except probably reduce the price of Vickers and Armstrong," Lieshman wrote Charles Schwab back in 1910. "Our own people you could trust, but I haven't so much confidence in Palmers, who were not in the original deal and in my opinion, has no better chance than N.Y.S.B. Co. [New York Shipbuilding Company], and may take advantage of low prices in bidding for Turkish ships, in order to force Vickers and Armstrong to place them in better position to bid on ships for the British Government."[121]

On April 8 members of the Union and Progress Party told Cabinet members that they would vote against them in Parliament unless the grand vizier provided to them the documentary evidence that the British Embassy continued to insist specifically for Armstrong subsequent to the Foreign Office communication of March 6. The grand vizier and the entire Cabinet (with one exception) came to Parliament with their documentary evidence, and they were closeted for two hours in private with the Union and Progress Party before the public session commenced. Since Palmers' success up till that point was due to its friendship with the Union and Progress Party, Talbot had drawn the inference that nothing would have prevented the entire party giving their votes to Palmers unless the grand vizier had produced the documentary evidence. Because these political secrets could not be mentioned in the public session, the grand vizier was left with little to say out in the open.[122]

In open session the Commission of Petitions asked for the examination of the file connected with the matter, but the naval minister refused. The assembly gathered to discuss the Armstrong-Palmers affair in the presence of cabinet members. The government reported that, in accordance with the program elaborated by the special commission under the inspection of Admiral Gamble, the government would purchase either two dreadnoughts of 19,250 tons with thirty cannons, twenty-one knots speed, or three battleships of 16,800 tons with eighteen cannons and twenty-one knots speed. Palmers had offered TL71.5 per ton and Armstrong more than TL100, but they had later reduced the price to TL86. The government then informed Palmers that they would supply the ships. After the withdrawal of the ex-naval minister, Halil Pasha, the subject of discussions with Armstrong was taken up without taking into consideration the offers from Palmers. The new naval minister, Mahmud Muktar Bey, after having studied the question, decided to separate the supply of the hull and the engines from that of cannon (armament). For the hull and engines Armstrong has asked TL51 against TL62 paid by England for the same work. The government had then arrived at an accord in giving the order to Armstrong for the price of TL51. In the words of Hakki Pasha, "Battleships are not bought like vegetables. We must have perfect units and coming from the best manufactories. This is why we do not desire to cut down prices too much and to assume a responsibility for the quality of the battleships."[123]

The Cabinet in the end became so nervous that members were compelled to submit to a vote of confidence as an alternative to resignation. The Parliamentary Committee of Investigation had reported to Parliament in favor of Palmers and against the Cabinet. The full Cabinet then presented themselves before Parliament. The grand vizier stated that the entire Cabinet would resign unless Parliament accepted its decision for Armstrong. The grand vizier stated openly during the public session that the Turkish Admiralty had chosen Palmers, but that the Cabinet decided for Armstrong after an exchange of views between the Ministry of Foreign Affairs and British Foreign Office.[124] In all, 115 Union and Progress members voted for the Cabinet and Armstrong and 40 members voted for Palmers with about 150 members abstaining. Talbot recounted how he had gone down to Parliament with the U.S. ambassador and they were present throughout the debate, "which ended with scores of Members banging their desks and shouting and shrieking at the Ministers to produce the technical designer from the Admiralty."[125] In the end, the Turks opted for Armstrong and Vickers as their battleship suppliers. The final contract signing with Armstrong took place July 27, 1911.[126]

Vickers gained even more by offering a particularly sweet deal to the Turks. In July 1911, Vickers' bankers Glyn Mills guaranteed a £600,000 advance to the Turks in respect of payment due to Vickers on the warships. Vickers provided for six months free credit, payment in ten equal parts, and paid for the start-up costs. Consequently, Vickers became the primary naval supplier in the final phase of Ottoman policy. In 1911 Vickers had contracts from the empire worth approximately TL2.2 million, and promises for a further TL5.5 million in 1913.[127] As a stopgap measure, the Turks bought the *Rio de Janeiro,* which Brazil had intended to buy from Britain. The ship, renamed *Sultan Osman,* was priced at £3.4 million, but the Ottomans bought it for £2.3 million in 1913.[128] When Djemal took over as naval minister in February 1914, the dreadnought *Sultan Osman* had already been bought from Brazil, and *Reshadiye* was still under construction in Britain. The Turkish government imminently expected war with the Greeks. Therefore, the government considered fleet reorganization a vital matter, and Djemal saw as his main goal taking delivery of the two dreadnoughts from their shipyards in England.[129] Ultimately, this British battleship deal with Vickers turned out to be unfortunate for the Ottomans. After the Turks had paid the last installment for the *Sultan Osman* in early August 1914, the British government commandeered both the *Sultan Osman* and *Reshadiye* to augment the Royal Navy after the outbreak of World War I.[130]

The Turks also purchased ships from the French. On April 30, 1914, the Porte agreed to a contract with Schneider for the construction of two submarines costing 2.2 million francs each and six destroyers from Forges et Chantiers de la Mediterranée. On May 2, 1914, the French shipbuilder Normand won construction of six torpedo gunboats and a subsequent agreement for twelve more. At the same time the St. Nazaire and Le Havre shipyards were set up for seven gunboats for the Ottomans.[131] According to Naval Minister Djemal, the French wanted to build ships for the Turks, and despite the disadvantage of having warships of different types in the fleet, the Turks accepted the offers "in order to please the French."[132]

The Young Turks did not limit their ambitions to upgrading the class of their navy. They also sought to strengthen domestic production by developing the capacity to construct their own dreadnoughts. They pursued a naval docks contract that would put in place the necessary infrastructure for self-sustained warship construction within the empire. The proposal for a dockyard reorganization originated with Armstrong. Armstrong sent a letter to the Turkish naval minister in November 1908, expressing its willingness to place at the disposal of Turkish government all their resources for the construction of ships, torpedoes, guns, ammunition, and floating docks. Additionally, "they would also be disposed under fair conditions to undertake the entire reorganization and equipment of the Imperial Naval Yard at Constantinople and if they were entrusted with this work, they would associate with themselves the house of Vickers Son and Maxim and probably another leading English firm."[133] The British government instructed Lowther to give his support to Armstrong-Vickers.[134] By early February 1910 a combination of Vickers, John Brown, and Armstrong had been arranged for Turkey.[135]

The firms' crowning achievement in 1913 was the acquisition of the docks contract. Here Vickers, Armstrong, and John Brown collaborated to establish the Imperial Ottoman Docks, Arsenals and Naval Construction Company for the exploitation of Ottoman state docks and arsenals. The capital of this enterprise was £250,000, of which the Ottoman government held the controlling share, and the minority was divided between Vickers and Armstrong. Armstrong-Vickers contracted an Ottoman loan for TL1.485 million at 5.5 percent interest.[136] The chief conditions of the proposal committed the British firms to invest £1.2 million. In exchange, the Ottoman government guaranteed a sum of £84,000 annually as interest and sinking fund on the British investment. Furthermore, the Ottomans committed to entrust all of the new naval program construction to this group immediately, and to bind themselves for thirty years to place all naval work with the group.[137]

In pursuing the Dock Commission with the British, the Turks provoked German opposition. In October 1913, the Ottoman government planned to reach the agreement with the British consortium about the dock construction at Izmit and the transfer of all Ottoman naval orders to Britain. The German representative in Constantinople wrote to Berlin that Germany could not allow this, and argued to the Ottomans that the late Mahmud Shevket Pasha had promised him to address orders concerning battleship construction to Germany. Nevertheless, Germany would approve the deal with Armstrong if the Ottomans would buy a £500,000 dreadnought from Germany. In response, Armstrong threatened to refuse to construct the dock or arsenals if Germany's demand was not refused. The matter was discussed in the Ottoman Council of Ministers, which endorsed the concession with the British firms. In November the Turks offered to place immediately £4 million of naval orders on the terms of an annual cash payment in the amount of £1 million for four years, contingent on an outside loan. Finally, in July 1914 the Ottoman government and the British firms approved the joint agreement, and the documents were ordered sealed.[138] Thanks to the dockyard concession Naval Minister Djemal Pasha had "the great pleasure of knowing that within a short time we should find ourselves in possession of an arsenal, building yards, harbours and factories of the latest pattern."[139]

In the wake of this triumph, British firms pressed on for more. By 1914 Vickers, Armstrong, and John Brown teamed up and secured Turkish orders for three dreadnoughts, six destroyers, two light cruisers, and two submarines.[140] Joining forces, the three firms held a tremendous advantage because they offered a unified package of hulls, armor, and armaments. Additionally, British armament products were also competitively priced and British producers could offer rapid delivery (twenty-four months for the largest-class warship, while German producers were 30 percent slower).[141] Once again the elements of expediency and money manifested themselves in Ottoman armament policy. As Djemal Pasha, the Young Turk minister of marine, recalled in his memoirs, "In view of all this activity, it will at one be admitted that our one object in life was to make our fleet superior to the Greek fleet at the first possible moment. I did everything conceivable to remove all obstacles and prevent any delay in the realization of this project."[142]

The Greek naval race against the Turks escalated up to the Balkan wars and beyond. Following the Turkish dreadnought purchases in 1911, the Greeks called for tenders for a battle cruiser in 1912. Based on the lowest price the Turks granted the contract to Vulcan for the *Salamis*.[143] As Paul Halpern noted, by January 1914 the naval business competition had heated up to such a degree that the French viewed the British, not the Germans, as their chief rival.[144] Desperate to match the Turks, in May 1914 the Greeks bought two older battleships from the United States. Greece paid $12.5 million for these outdated battleships, renamed *Lemnos* and *Kilkis,* and the warships joined the Greek navy in July.[145]

The disastrous performance in the Russo-Japanese War coupled with the revolutionary upheaval in 1905 ended formal autocracy in tsarist Russia. The new legislative body, the Duma, could approve or reject the annual budget. Nevertheless, the procurement process did not change much from the pre-1905 era. The Ministries of War and Navy still largely controlled the granting of contracts, and, as previously, the government continued its policy of promoting domestic production. In 1907–1910 military orders placed abroad decreased from 16 percent of military outlays to 4.9 percent. A similar trend occurred in naval orders, with the foreign share dropping from 36 percent in 1907 to 23 percent in 1909. After 1910 the government encouraged domestic firms that employed foreign partners, and formal armament imports largely disappeared. Instead, foreign armament firms set up plants within Russia or participated in Russian companies through direct investment.[146]

The potentially enormous Russian naval orders drew the immediate interest of private foreign suppliers. Assessing the opportunities for exports to Russia in 1912, Sir Charles Ottley, a board member of the British firm Armstrong, observed:

> Russia is the only remaining European Power of this magnitude who is not self-supporting in her war material. She alone therefore of the Great Powers can offer fair prospects of business of many years to come. She is a vast untilled field for industrial enterprise, and I do not like to turn our backs upon this field until we have of all events considered the risks and looked at the possibilities calmly in the face.[147]

The Naval Ministry allowed Vickers, Armstrong, Schneider, and Krupp to tender for thirty fourteen-inch, forty eight-inch, and one hundred six-inch guns. The total order was valued at 25 million rubles, but the navy insisted that the guns be manufactured in Russia and that they be completed in January 1916.[148] In addition to the naval gun contracts, the Duma's allocation of 470 million rubles for the first installment of the naval program sent foreign firms scurrying to find Russian business partners. As Ottley noted, "Our friends in Paris (Schneider) are interested in Franco-Russian, Nevskii, and Putilov, John Brown interested in Baltic yard and new Nikolaev, Hamburg (Blohm und Voss) will also build some turbine machinery."[149]

The prospect of such vast sales intensified the rivalry and intrigues of the foreign competitors. When General N. V. Ivanov, chairman of Nikolaev Shipbuilding Company, called at the Admiralty for an appointment with Assistant Minister of Marine Admiral M. V. Bubnov, Bubnov chastised the general for coming to him on foot, "and chaffingly enquired what sort of chance he expected to have compared with that of rivals who drove to the Admiralty in Imperial carriages and upon whose behalf Grand Duchesses personally telephoned shortly before their arrival asking that they might be received without delay." The comments were in reference to Eugene Schneider, who was staying with Grand Duchess Maria Pavlovna. Schneider had paid 3 million francs to Maria Pavlovna for "erecting salt works on H.I.H.'s estate."[150] Later, General Ivanov informed the tsar about Schneider's maneuvers, and Nicholas told the grand duchess to convey to Schneider that "he quit the country within 24 hours." Schneider left St. Petersburg March 25, and Creusot was thus eliminated from consideration for the new gun factory.[151]

Although Schneider may have lost that battle, he did not lose the war for Russian business. Schneider-Creusot succeeded as the first foreign manufacturer to receive a contract for the construction of a factory and dockyard within Russia. The contract began October 29, 1912, and also included the Russo-Baltic Works and the Revel factory. Under the terms of the agreement, Schneider guaranteed the production of the ship bodies for destroyers and submarines, while the German firm Vulcan manufactured the engines and aided in turbine construction. For its part Schneider would receive a percentage profit per destroyer and submarine in addition to 500,000 rubles annually.[152]

Russia's naval rearmament after 1908 felt the effects of the Greco-Turkish naval race because the Turkish jockeying for naval advantage with Greece caused growing alarm for the Russians. Since Russia lacked modern shipyards on the Black Sea and Russian warships could not enter or exit through the Turkish straits, there would be no way to counter Turkish dreadnoughts in the theater. In response to the growing threat, the Duma passed a Black Sea building program that Tsar Nicholas II signed into law in 1911. This program called for reconstruction and modernization of the shipyards at Nikolaev, where three dreadnoughts were to be built.[153]

For the modernization of the Black Sea fleet, the Russian government turned to Vickers. In 1911, Vickers established the construction of Vickers-pattern warships at Nikolaev by putting up 10 percent of the capital. Vickers then secured 20 percent of the holdings of the Tsaritsyn naval gun foundry in 1912. The next year, Vickers took an even greater share of the Tsaritsyn project. In addition to pocketing a 3-million-ruble design fee, Vickers obtained 10 percent of the gun-making profits over a guaranteed period of ten years.[154]

Vickers' position in the Russian defense market did not appear above reproach. One of Vickers' shipbuilding partners in south Russia carelessly made a formal entry in the company accounts in the amount of 100,000 rubles "expended at the Ministry of Marine." Although it remained unknown who received the money at the admiralty, the intermediary was reportedly Basil Zakharov, a member of the Vickers board of directors. The incident caused the British ambassador to observe, "In this connection it may be remarked that comment is not infrequently made here upon the seemingly rather favoured situation extended by the Russian Admiralty to Messrs. Vickers, especially as the delay with and unsatisfactory condition upon delivery of the armoured cruiser Riurik have left disagreeable impressions in many quarters."[155]

The regional naval races in the dreadnought era had wide ramifications. Although each of the races had a bilateral rivalry at their core, the dreadnought races evolved into three-way affairs. In the case of South America the Brazilian-Argentine race eventually dragged in Chile as a third party. Similarly, in the Eastern Mediterranean the Greco-Turkish tensions operated as the motor for the race, but Russia felt the repercussions across the Black Sea. Even though the Turks placed their emphasis on overtaking the Greeks, every Turkish acquisition changed the equation for Russian naval power on the Black Sea. Although Russian naval allocations generally gave priority to the Baltic fleet, the Black Sea drew more attention than it otherwise would have because of the Turkish moves. Precisely because the force sizes were smaller on the Black Sea, the addition of one or two dreadnoughts could dramatically alter the strategic situation.

The growing intensity of the naval business compelled the companies to set up shop within foreign countries or run the risk of being excluded from lucrative markets. As the examples of Schneider and Vickers make clear, the Russian tsarist government preferred as much as possible to procure warships from domestic sources, and so exporters found themselves forced to turn into Russian producers by establishing facilities in country. The Turks also were striving to enter the ranks of the dreadnought builders, and they looked to Vickers to provide them with those capabilities. However, in the short term the Young Turks hastened to purchase ships outright while the Greeks also pursued naval expansion with a conscious aim of achieving superiority over the Turks through imports. In contrast, the South American customers remained principally importers.

The dreadnought races in South America and the Eastern Mediterranean cast serious doubt on the notion that democratization reduces armed competition and automatically promotes peace. The South American republics engaged in a second round of naval races with full backing from the national legislatures. In the Ottoman Empire, the overthrow of Sultan Abdulhamid's autocracy and the introduction of a constitutional order requiring parliamentary approval of naval procurements did not bring a reduction in naval acquisitions or a decreasing likelihood of embarking on armed competition. Just the opposite occurred. The introduction of even a limited democracy into the procurement process made possible the employment of naval acquisitions as a sign of popular will and political leadership, thereby transforming the arms trade into an arms race. The increased political pressure on the procurement process helps to explain why an incident such as the Palmers affair could ignite a political firestorm and potentially bring down the Young Turk Cabinet. The Young Turks viewed naval expansion as a means of Turkish national revitalization, not simply a matter of defense.

7

Gunning for Krupp

In the decade preceding the outbreak of World War I all the states in Eastern Europe and South America undertook major rearmament of their land forces. In technological terms the common process of artillery modernization throughout the regions involved the acquisition of the improved quick-firing technology for field guns and mountain artillery. In business terms, Krupp's deeply entrenched position came under fire from its chief foreign rival, Schneider-Creusot. Schneider, with powerful backing from French banking, executed a concerted program to sweep the Germans from the field. Krupp also faced competition from the fellow German firm Ehrhardt (RheinMetall) and also Skoda of Austria. In Eastern Europe Krupp prevailed in Turkey and Romania, but faced defeat in Bulgaria, Serbia, and Greece. Schneider's ascendancy culminated dramatically in the First Balkan War (1912–1913) when the small Balkan states, armed with French artillery, overwhelmed the Turks and their Krupp guns. In South America, despite poor performance by its hardware compared to its rivals, Krupp preserved its monopoly position, thanks to the favoritism of Germanophile army officers and blatant corruption of the procurement process. French threats of financial retaliation against the buyer states did not intimidate the South American customers sufficiently for them to buy the better Schneider gun.

By the turn of the century various rifle and cartridge manufacturing firms, mostly German ones, coordinated competition for sales and arranged production sharing through the Deutsches Waffen- und-Munitionsfabriken (DWMF). In November 1896, DWMF was founded through the merger of the following firms: Deutsche Metallpatronenfabrik (Karlsruhe), Loewe (Berlin), Rheinisch-Westfaelischen Powder Company (Cologne), Rottweil-Hamburg Powder Company (Rottweil), Fabrique Nationale d'Armes de Guerre (Herstal, Belgium), and Mauser (Oberndorf). Steyr entered into the arrangements a few years later. In 1905 all the firms formed a cartel by agreeing to submit uniform bids on foreign orders and then divide the work among themselves. Each firm would tender a minimum price of 75 francs per rifle, of which 15 francs would go into a common fund, with Steyr taking 37.5 percent and DWMF 62.5 percent. As part of these arrangements DWMF waived its exports to Greece and Serbia in favor of Steyr in 1905, clearing the way for Steyr to contract with the Serbs to provide 30,000 Mauser rifles in 1907. As an extension of the agreement Steyr refrained from competing in Turkey or Spain in return for DWMF's ceding Bulgaria and Romania to the Austrian firm. In effect, Steyr and Mauser agreed to defined spheres of influence in the Balkans and divided the firearms market between themselves until 1914.[1]

The technical advances in quick-firing guns drove the artillery trade in the first decade of the twentieth century. France had taken the lead with its 75mm recoil field gun in 1897, and other countries eagerly sought to modernize their equipment and rearm their armies. The technological advantage gave Schneider-Creusot a weapon with which to challenge Krupp's long-standing dominance in the market. Schneider competed against Krupp most intensely in the Balkans and South America. The value of Schneider's arms exports rose sharply, from 35 million francs in 1898–1899 to 85 million francs in 1907, and then reached the pinnacle of 140 million francs during the Balkan wars, 1912–1913. In November 1912 Balkan orders amounted to 127 million francs, and during 1913 out of Schneider's 170 million francs in artillery and naval orders, Balkan countries accounted for 60 million.[2]

At the same time the Skoda works in Plzen followed Krupp and Schneider into the export markets, although with less success. Like the Germans and the French, the Austrians saw the Balkan states after 1904 as the most promising market for growth. Romania and Serbia showed a willingness to consider Skoda guns, and the Serbs received a sample gun and 1,000 shells for trials. Skoda anticipated a Serbian order, but the coup against the Serbian ruling dynasty in 1903 brought in a new government that decided to conduct a new series of trials. Skoda found limited success with sales to the Ottoman Empire. In 1913 Turkish purchases helped contribute to an economic upswing for the firm, as the order accounted for roughly one-sixth of Skoda's 36 million crowns' worth of foreign arms sales in 1913. Nevertheless, Skoda appeared distinctly behind Krupp and Schneider in the second ranks for artillery exports.[3]

After the Russo-Japanese War, China pursued rearmament and modern artillery. In April 1907 at Tientsin the government held gun trials for field and mountain guns, including Skoda, Krupp, and Schneider, but no order followed. In September 1910 Skoda concluded an order with the army of Heiklungkiang for eighteen quick-fire artillery pieces (75mm) at a cost of 19,170 marks per piece. This breakthrough into a Krupp-controlled market occurred because of a bit of Chinese intrigue, since the Skoda gun was a cheap copy of the Krupp gun. In 1911 Skoda sold six machine guns and field guns to China. Skoda further offered financial support to the Chinese government through a £150,000 loan at 6 percent interest to purchase war supplies from Skoda. Following the fall of the Manchu government and the establishment of the Chinese National Republic, the contract was cancelled and replaced with a new loan for £450,000 from Austrian banks to pay for Skoda guns, Hirtenberger cartridges, and Steyr rifles in January 1912. These items were delivered to China in 1913–1914.[4]

German suppliers continued to dominate in Chinese military sales even after the fall of the Manchu dynasty. From the 1890s to 1911 the Germans sold 25 million to 30 million marks' worth of armaments to China. Moreover, German firms secured 65 percent of foreign armaments orders to China in 1909–1911. These weapons accounted for 5 percent of all German exports to China. From mid-1912 to mid-1914, China imported 91 percent of its weapons from Prussian depots, for a total of 260,000 Mausers (model 1888).[5]

After its victory over Russia, Japanese arms imports changed in scale and source as domestic capabilities increased. Between 1904 and 1908 Japan bought only 200 Krupp pieces, and from 1908 until 1911 only 151. The Austrians emerged as a new supplier in this period. In 1908 Steyr sent two machine guns to Japan, and in 1912 Steyr sold twenty machine guns to the Japanese navy. Meanwhile, in 1911 Steyr had delivered 500,000 rifle barrels to the Japanese army. In 1914 Steyr exported an additional twenty-four machine guns and 8,000 cartridges to Japan to be adapted for Japanese use. The Austrian ambassador in Tokyo crowed that Austrian sales had come to exceed English naval sales and German artillery sales to Japan.[6] While this was a true statement, the relative magnitude of Austrian sales more accurately reflected the decreasing role of foreign arms purchases for Japan.

By 1903, with the Bulgarian government contemplating rearming its artillery, Schneider and Krupp were positioning themselves for an intense competition for dominance. The Germans had an advantage in that the Krupp system had been the standard in the Bulgarian service. Therefore, all the current shells could be used only in Krupp guns. The inspector of artillery was talking to Krupp representatives in Sofia about quick-fire guns, and a Bulgarian captain in Berlin was pushing Krupp's system. Even though Bulgaria possessed Schneider guns delivered in 1901 and 1902, it had no shells in store. If Schneider hoped to gain ground, it would have to obtain some stake in the contracts for shells. The Bulgarian minister of war, Mihail Savov, and the Council of Ministers recognized the need for Schneider projectiles in their mix. In July 1903, the Bulgarians judiciously split their shell orders. From a total order worth more than 3 million francs, one-third was awarded to Schneider and the remaining two-thirds stayed in Krupp's hands. Negotiations between the War Ministry and Schneider had been going on for some time for the purchase of seventy-two quick-firing guns made originally for the Boers at the Creusot works. However, because these guns varied in model and required considerable alteration, Bulgaria decided not to purchase them.[7]

Schneider seemed poised to score its first major victory over Krupp. When the Sobranie opened its session the legislature approved an "exceptionally heavy military expenditure" for an extraordinary credit of 25 million francs to purchase equipment, and an additional 7 million francs (4.5 million for past deficits, 2.5 million to purchase 15,000 rifles from Steyr).[8] Rumors were already circulating that Bulgaria might buy Creusot 7.5cm guns (twelve batteries). Bulgaria had taken delivery of eighty field howitzers the previous year, and the projectiles were loaded with Schneider's recently invented explosives, which were the most powerful type at the time. Schneider's prospects looked promising. In the estimation of the British consul, "It is Le Creusot works which chiefly furnish this country today, not Krupp which has fallen into disrepute. Messrs. Schneider and Co. of Le Creusot offer great facilities for payment, and this government may be thus enabled to give orders, putting off the evil day when payment becomes necessary." As an example of the credit given, the French had overlooked payment of 5,000 francs owed for three years. When Bulgaria at last made the payment, Schneider had not demanded it, and the firm issued the receipt without calling attention to the matter.[9]

Although Creusot appeared set to claim the order for field guns, financial matters intervened. With embarrassment, M. Revol, one of Schneider's directors, reported back in late February that the business found itself in an "imbroglio," and negotiations had ground to a halt, chiefly owing to financial questions. The ministry also opposed the potential distribution of the order. Revol blamed the troubles on a poor informant. By March, Bulgarian commissioners were visiting Germany and other countries to buy guns. Financial concerns figured prominently, as most military furnishers required some solid guarantee of payment, which Bulgaria did not possess.[10]

Bulgarian decisions in April 1904 brought mixed results. The country concluded an order with Schneider for three torpedo boats. Paul Pichon, the French naval officer directing the Bulgarian flotilla, had been instrumental in securing the order. However, the field gun question was still hanging. Much to the dissatisfaction of the French, Bulgaria bought 50,000kgs of gunpowder from Rottweiler Pulver. The Creusot agent was annoyed by this order because a clause in the loan agreement for £1 million in France stipulated that preference should go to French firms. Even more galling, the War Ministry purchased fifty-four mountain guns from Krupp, and deliveries were to start in August. After an examination of Krupp's quick-fire field guns, the Bulgarian inspectors concluded that the guns had defects in the laying mechanism. In all likelihood, the complaint was an excuse to delay making the decision on which gun to purchase. The real reason was the need to raise further Bulgarian loans in France unless the order for quick-firing artillery was placed with a French firm.[11] The army would acquire no quick-fire field guns until the government resolved the financial matter.

Afraid that it might have pushed the Bulgarians too hard, Schneider discussed a new approach with the French banks. In September 1904 the firm convinced the Banque de Paris et Banque de Pays Bas (Paribas) to conduct the negotiations of a loan with the Bulgarian government and to offer them a proposition more readily acceptable than the last one. Contrary to the advice of his colleagues at the Paribas, president Edouard Noetzlin believed the dangerous German financial competition had caused the possible intervention of the German emperor and Krupp. He promised to make every effort to get the bank to send a delegate to Sofia bearing an acceptable financial proposition, giving the Bulgarian government the possibility to receive liquid funds as needed at the time of contract signing. Schneider hoped to get the Bulgarian government to accept a loan of 100 million, with 80 million firm and 20 million on option, with the latter sum destined for railroad enterprises. On the other side, Revol had met with L. Renouard, vice president of the Council of Paribas, who had insisted on the opportunity to prevent the Bulgarian government from turning to German banks. Schneider also conveyed to the bankers the French foreign minister Theophile Delcassé's desire to see that the Paribas take back the negotiations with Bulgaria on an acceptable basis. Renouard promised to submit the question anew to his colleagues. As an incentive to bring the Paribas on board, Schneider suggested allowing them to augment their rates of commission.[12]

By late October Revol had arrived in Sofia and delivered to Henri Bousquet the response by Paribas to requests presented by the Bulgarian government regarding the first project of the contract. Bousquet was authorized to sign the definitive contract in the name of the financial consortium. The response of the Paribas satisfied all the demands of the Bulgarian government. The bank's chief motive for concessions was fear that the German government had not ceased telling the Bulgarian government that Bulgaria would find in Germany better terms than in France. If Bousquet had continued to hesitate, the German group (Dresdner Bank and Deutsche Bank) and the Austrian (Banque de Commerce de BudaPest) would have proposed 100 million francs to the Bulgarian government without the annoying condition concerning railroads and with the sole obligation to order war material in Essen. Certain that Krupp had made desperate efforts to keep the Bulgarian order, whose loss could compromise orders in Greece and Serbia, General Racho Petrov had insisted on concluding the deal rapidly, and he had confirmed the intervention of the German government through its minister in Sofia.[13]

Three German banking groups competed for the Bulgarian loan: Deutsche Bank and Banque de Commerce de BudaPest; Groupe Disconte Gesellschaft (Berlin) and Wiener Union-Bank (Austria); and Dresdner Bank, officially through the German government. The first two groups proposed similar plans, offering Bulgaria 60 million with absolute freedom to order its artillery in France or elsewhere. Dresdner offered 40 million on the condition that artillery be ordered in Essen. Revol, reporting from Sofia, worried that the German banks were looking for a way to obstruct and to bring about a rupture with Paribas by trying to persuade the Bulgarian government that it should demand conditions better than those obtained in 1902, conditions that the German banks were disposed to underwrite. Once the rupture had been effected with the French consortium, the Bulgarian government would find itself facing a syndicate formed by the reunification of the three German groups. The outcome would have ramifications throughout the region. As Revol remarked:

> This is not purely a question of material orders, the German Government has sustained the struggle up to now with a simple political goal, and its importance is considerable. The new issuance made in France will be consecration of the financial independence of Bulgaria vis-à-vis the Austro-German grouping and it will constitute a serious failure for German influence in the Balkans, influence which exists mostly now in Turkey and Romania and which seeks to penetrate into Bulgaria, and it rests on two principal factors: war materials and financial assistance.[14]

Schneider's financial allies brought the firm victory. On November 13, 1904, Bulgaria officially signed the artillery contract with Schneider for eighty-one batteries of 75mm quick-fire field guns, with each battery consisting of four guns and twelve caissons. Bulgaria made payment in the form of state treasury bonds bearing 6 percent interest for thirty-six months. As a reward for the successful outcome, the participants enjoyed hefty commissions. The total commission represented 3.5 percent on Schneider's net price. Out of that amount, Banque de Paris agreed to a 5 percent share; Bulgarian officers Balabanov, Savov, and Petrov accepted 10 percent; and 2 percent was divided among four ministers. The nominal value of the Paribas loan amounted to £4 million, out of which Paribas retained £2 million to meet the charges on the purchase of the Creusot guns and construction of the Trans-Balkanic Railroad. The 324 Schneider guns cost £1,027,080. The keen competition from Krupp had driven down Schneider's final price to 317,0000 francs (£12,680) per battery, and this price was considered low by observers.[15]

To continue to acquire armaments the government sought a new Bulgarian state loan with Paribas in February 1907 to convert the loans of 1888 and 1889 at an interest rate lower than the 6 percent they carried. For the new loan French financing played the key role, although German banks (Disconto Gesellschaft, Deutsche Bank, Bleichröder) and Austrian banks (Länderbank, Bank Verein) also participated. The French did not require a guarantee for this loan, and they offered terms more advantageous to Bulgaria than the loans of 1902 and 1904. After conversion, the bulk of the loan, nominally 145 million francs (£5.8 million), was earmarked for military expenditure.[16] Clearly, the French bankers were sweetening the pot in anticipation of locking up the artillery orders for Schneider.

In March 1907, the Artillery Commission spoke out against concluding the contract with Creusot for 21 million francs. The commission's decision was conveyed to the Council of Ministers by General Savov, and the Council decided to pass it to the parliament for examination. In this way the ministers got out of a tough position. The French government remained true to its own business by letting the Bulgarian government know that until the contract was concluded there would be no loan on the Paris Bourse. According to the Commission, the difference in price between Schneider and Krupp for the whole order was 1,528,930 francs. Under the terms of the new order with Creusot, the French firm charged Bulgaria 290 francs for 15cm shells, while Krupp asked only 150 francs. Similarly, for an order of 118,000 quick-fire shells for field guns, Schneider was charging 62 francs per shell. The government's comparisons revealed that the Russian government had paid Schneider only 54 francs per shell during the Russo-Japanese War, and Ehrhardt asked only 45 francs per shell.[17] The Germans received encouragement from Bulgarian foreign minister Dimitur Stanciov, who declared to the German ambassador that "the guns would definitely not be ordered from Creusot, but rather would be placed with Krupp."[18]

The Sobranie trumped the Artillery Commission and approved the sale with Schneider. They voted a special credit of 32 million francs, mostly to be spent in France. Schneider received 25 million francs in artillery orders, and German and Austrian sales claimed part of the remainder. Bulgarian legislators understood that the loan contract with French banks obligated them to place the bulk of the orders in France. Out of the £5.8 million loan, military orders claimed £1.28 million, of which Schneider took £1 million for thirty-six quick-fire mountain guns (7.5cm), thirty-six quick-fire howitzers (10.5cm), ammunition, torpedoes, telegraphs, and telephones. German and Austrian firms also gained modest orders for firearms and ammunition: £112,800 to Roth of

Vienna for 32 million rounds of small-arms ammunition; £40,000 to Rottweil for 90 tons of smokeless powder; £80,640 to Berlin Waffen Fabrik for 144 Vickers-Maxim machine guns; and £19,200 to Herr Buddenbrook for 10,000 Mannlicher rifles.[19]

British firms proved unable to break into the Bulgarian market. Near the end of the year Colonel John Du Cane met with Colonel Elias Dmitriev, the chief administrator of the War Ministry. Dmitriev had recently met with Armstrong's representative, Kaufmann. Dmitriev believed that a British firm such as Armstrong should receive an order for war material from his government. Unfortunately he could not overlook the difficulties in connection with transport charges and the financial aspect that would confront Bulgaria in giving such orders. Dmitriev explained the secret arrangements made by the Bulgarian government with the firm Continental Transport Agents. That firm charged only one-sixth the price paid for transport across Europe of war material supplied by other firms. Dmitriev made it clear to Du Cane that unless the British firm employed Continental Transport Agents, the expense of bringing British goods across Europe to Bulgaria would force the government to give the order to a continental firm. If the British goods could come by sea through the Turkish straits, however, the costs "would be so small that the Bulgarian Government would certainly be thereby almost irresistibly attracted towards dealing with a British firm. He suggested that, considering the fact that gun-boats supplied by a German firm have recently been allowed to proceed through the Bosporus for delivery to the Romanian navy, the existing obstacle of Turkish interference with consignments of war material for Bulgaria may not be quite insuperable."[20]

As an example of the intimate connection between the negotiation of Bulgarian loans and the purchase of war material, Dmitriev pointed to the negotiations for the most recent loan (concluded in March 1907). Discussing the financial aspect, Dmitriev explained that "in the present stage of Bulgarian national and industrial development no loan for railway construction or similar reproductive services can be negotiated with any foreign financial house unless it be encumbered with conditions involving the purchase by the Bulgarian Government of war material or other industrial products of the country in which the financial house furnishing the required capital operates." The preference that the Bulgarian Ministry of War had for Krupp could not be followed because the

German financial houses declined to provide capital. The government was therefore forced, primarily to raise the money for the conversion of two of their loans, to turn to a French financial group that was prepared to negotiate the loan on condition that Schneider should be given an order for 25 million francs (£1 million) worth of war material. According to Dmitriev, none of the parties, except the French, was completely satisfied, and the Germans, in particular, regretted the proceedings, which had put them in the background of the Bulgarian market. Dmitriev offered the conclusion that unless British manufacturers included offers of general financial support with their proposals for war orders, they could not expect to receive the consideration "of which no small number of Bulgarian experts believe them to be worthier than many of the foreign manufacturers now patronized." Finally, Dmitriev raised his main point. A British arsenal in Bulgaria would avoid the transport and financial aspects, and Dmitriev personally favored such a solution.[21]

Dmitriev's comments were true, but one can discern in them a calculated attempt to manipulate British policy. British firms did face a disadvantage due to a lack of financial support. By disingenuously casting Bulgaria as the reluctant recipient of war orders, Dmitriev painted a picture of a poor, small country being forced to accept guns in order to gain needed loans for peaceful, productive economic purposes. No doubt the Bulgarian officer hoped to play on well-known British antipathy to arms sales in the region as a waste of money. The Bulgarian objective was to obtain British diplomatic help in easing restrictions on shipping its war materials through the Turkish straits by dangling the prospect of purchases from Britain. Failing that, perhaps Bulgaria could entice Armstrong to invest in Bulgarian home production. The British Foreign Office did not warm to the prospect. The margin comments on the future requirements of the Bulgarian army with regard to war material noted, "suggestion that Turkey might not object to the passage of ships for Bulgaria through the Bosporus is not to be entertained," and, with regard to Russian relations, "It would expose this country to great suspicion if a British Arsenal were established in Bulgaria."[22]

Bulgaria never did succeed in finding a foreign sponsor for its domestic plant. In March 1908 a rumor circulated that Krupp was proposing to build a shell and gun factory in Bulgaria under condition that all orders be allocated solely to that factory. The Bulgarian war minister told the Russians that only Armstrong had proposed to build a powder factory in

Bulgaria. According to Russian military attaché Leontiev, Krupp's agent had likely started the rumor.[23] In the end, nothing came of it.

Schneider had triumphed in Bulgaria thanks to superior technology and muscular financial support. As Dr. Stojan Petrov Danev, president of the Bulgarian Parliament (Sobranie), explained to Lord Stuart Rendel of Armstrong in 1912, "the reason that the orders were placed with Schneider was that the issue of Bulgarian loans had usually been effected through French houses, who made one of their conditions the ordering of war material through French firms. Confidentially he said that French artillery for Bulgaria was thoroughly satisfactory but very expensive."[24]

For the French, Bulgaria served merely as the all-important first step in a concerted effort to roll back Krupp throughout the Balkans. After concluding its first major artillery contract in 1904, the French military attaché in Sofia confided to his Russian colleague that the victory of French industry through the Bulgarian order practically guaranteed the next order for 108 batteries over the next two to three years. More important, French success in Bulgaria could "influence the orders of Serbs, Greeks, Spanish, Belgians, and even the Turks."[25] The Germans viewed the situation in the Balkans the same way. As they tallied up Creusot's 50 million francs' worth of artillery orders between 1903 and 1907, Berlin worried that the French influence was replacing that of Austria and Germany, and was turning Bulgaria into a "French colony."[26] The stage was now set for a renewal of the Krupp-Schneider rivalry in neighboring Serbia.

At the beginning of 1903 Austrian firms were poised for success in armaments orders to Serbia. In the first week of January Vienna heard the good news that Austro-Hungarian industry had beaten German, French, and Belgian munitions firms for Serbian cartridge orders. The firms Manfred Weiss (Budapest), Roth (Vienna), and Hirtenberger (Niederösterreich) had secured equal portions of a contract worth 5 million francs for 25 million to 45 million cartridges. In the process, they had beaten back attempts by Rottweil Pulver of Cologne to discredit them. The orders still needed approval from the Skupshina, where the radicals opposed the vote for supplemental credits for the military deliveries, but King Alexander's government supported the Austria orders.[27] At the same time, Skoda stood on the brink of a major export order. After nibbles from Belgium and Romania, Skoda was approached by the Serbs about adopting the Austrian gun for the rearmament of field artillery. Responding promptly to the Serbian interest, Skoda sent a test gun to Belgrade with promising results. In the directors' meeting in Plzen on July 18, 1903, the board heard news that the Serbs had almost settled the matter, and the company anticipated the Serbian war order in October.[28] Indeed, on May 13, 1903, Walter Trappen, Skoda's general director, informed the Austrian military attaché in Belgrade, Major Josef Pomiankowski, that the Serbs had offered the gun contract to Skoda. Trappen had telegraphed Serbian war minister Petrov with the news that Skoda had accepted King Alexander's offer, and he urged Petrov to inform the king immediately. Petrov reported back that he had met with the king, and the order would follow. Trappen knew full well that in "lovely Belgrad all possible intrigues could be concocted in order to change the mind of the King," but Alexander had given verbal assurances four times that he wanted the artillery order given to Skoda.[29]

The coup in Belgrade in June 1903 profoundly changed the course of the arms sale for Skoda, Austria, and Serbia. Indeed, Skoda's fortune was so tightly bound up with the fate of the king that Trappen was the last civilian to speak with the king before the king's murder. The change in the Serb ruling dynasty brought with it a change in the pro-Austrian foreign policy. By late July the new Serbian government informed Skoda that because of lack of funds and the absence of the required legislative approval for the credits, the artillery order had been annulled. Additionally, the Skupshina experienced great difficulties in passing a budget for military credits because of the political factionalism. At first Skoda's directors remained optimistic. Late in January 1904 the board continued to expect the Serb order by February 2.[30]

During King Alexander's reign Serbia had resolved to rearm its field artillery with more advanced quick-fire guns. In the legislature, the major political parties had agreed to seek a new foreign loan for artillery modernization. Without a loan, it was unlikely that Serbia could proceed with rearmament. In the spring of 1904 a German and French banking syndicate proposed a loan between 30 million and 40 million francs; Austria was purposefully excluded. Strong political opposition to this loan brought about the fall of the sitting Cabinet, and Nicola Pašić formed a new government on November 27. Meanwhile, in October 1904 the Serb Field Artillery Commission confirmed to Creusot that it judged Schneider material superior to the models presented by competitors during the past year. The Serbian government assured Schneider that it still looked favorably toward the firm's material. At the request of the Serb war minister and finance minister, Schneider sent its proposal to furnish field guns and for payments over multiple years. Based on its latest information from Serbia, Schneider knew that the government was prepared to dispense 10 million to 12 million francs and was sure to find the necessary resources for a loan. Schneider was also aware of a German-Austrian-Belgian financial group that would seek to impose on the Serbs a condition that all field guns must be German.[31]

By December 3 the Serb officers on the commission had spoken out against Skoda and in favor of Schneider. However, the business was not yet closed, and Austria hoped that Pašić would be supportive of a combination of Skoda-Wiener Bankverein. While German financial houses conducted meetings with French partners, and the Serb finance minister visited Paris on an official invitation, Skoda sent Georg Günther, its new general director, to Belgrade to lobby for orders. By late December Vienna determined to throw its diplomatic weight behind Skoda and instructed its military attaché in Belgrade that no other foreign power should be permitted to compete with Austria for Serbian arms sales.[32]

By this time the question of a Serbian military loan was a daily subject in the press. By late December 1904 the Serb government sought a loan for 70 million francs. Proposals included an advance of 8 million francs by an Austrian bank at 8 percent, and 25 million francs for field and mountain guns, and roughly 2 million francs for 75,000 rifles. Four financial groups engaged in negotiations: a French-German group (Sté Financière d'Orient); Dresdner Bank, on condition that all orders went to German industry; an Austrian group, demanding part of the railroads; and an English group, seeking a financial role with no conditions. The Serbian government wanted to go with first group (Naville), which had given great service to Serbia in 1902. However, for domestic political reasons, the Serbian government could not give the Naville group the concession on railroads, and it could only give French industry an order for material worth 7 million to 8 million francs. Schneider considered arranging a combination that would remove the German element entirely from the Naville group and replace it with Austrians. Under such a scheme Schneider wanted all the artillery orders to be French; the Austrians would receive the railroad and rifle contracts. To make this happen, Schneider approached Banque de Paris, which had relations with Länder Bank in Vienna.[33]

The Austrian ambassador energetically proposed to make the loan and place the order in Vienna. Pašić said that the government had not yet decided where to conclude the loan, and was waiting to hear from Paris. He divulged to the Russians that he preferred to go with French capitalists for political reasons and to order a French gun, either Creusot or St. Chamond. If the terms of the loan in Paris were disagreeable, then the Serbian government would negotiate with Berlin and buy from Essen. The main Cabinet reported that a representative of Berlin financiers had gone to Paris hoping to conclude an understanding with a French syndicate to include German banks in the loan. Pašić was reluctant to choose the Skoda gun because of the trade negotiations with Austria. He feared that "having such a powerful weapon in their hands, the Vienna cabinet of course would use it in the interests of the loan and purchases." In an attempt to satisfy all the competitors, the idea was floated in Serbia that Austria would get an order to convert 85,000 old Koka-Mauser rifles at 2 million francs.[34]

In January 1905 the Artillery Commission agreed to conduct field tests with St. Chamond, Schneider, Krupp, Ehrhardt, and Skoda guns. Austria delayed the testing until April by refusing to allow the transit of the non-Austrian guns across its territory. The foreign firms were told by Austrian customs that their consignments were not properly packed.[35] In February 1905 Günther asked the Austrian military attaché in Belgrade, Josef Pomiankowski, to provide a list of the Serbian officers comprising the Artillery Commission. A few weeks later Günther assured Pomiankowski that Skoda would be able to underbid Schneider for shrapnel. On April 2 Günther arrived in Belgrade for a personal interview with Serb officials.[36] Skoda still had grounds for hope. Just a few days earlier Pašić had told the Austrian ambassador that Austria would maintain one-third of all Serb orders.[37] Then, in late April General Rauomir Putnik, the Serbian war minister, sent an official letter to Skoda notifying the firm that the Artillery Commission, as the competent authority to evaluate the quality of quick-fire guns for rearmament, had drawn up the list of best pieces. "We regret very much," Putnik wrote, "that the name of your firm was not found on the list of cannons."[38]

As a way to avoid alienating Austria, France, or Germany, Finance Minister Lazar Pacu negotiated a financial entente. French, Austrian, and German banks would participate in the loan in roughly comparable levels, with the French providing 40 percent, while the Germans and Austrian put up 30 percent each. Those countries would be granted proportional shares of the industrial orders. Under this plan, the German share would consist of shell orders to Krupp, whereas all the gun orders would go to France. For the loan, Schneider made certain that a clause was inserted specifying the sum of 19.857 million francs for payment to the firm for artillery and munitions. In May 1905 this consortium drew up a loan for 110 million francs, including 41 million francs for foreign orders, of which 22 million francs were designated for artillery. The loan's high interest provoked opposition in the Skupshina, and Pašić resigned. In the ensuing elections of July 1905, the Independent Radicals, who had opposed the loan, emerged victorious. They promptly scrapped the French loan in favor of a smaller Austrian loan of 70 million francs through Bankverein.[39]

Because of the changes in the Serbian Parliament and the consequent preference for Austrian financing, Skoda's prospects improved dramatically. The Liubomir Stoianovic Cabinet decided that comparative gun trials conducted in Serbia would be impossible, given Austrian disruptions. Therefore the government had the Artillery Commission visit the different factories to evaluate each gun system. Although the artillery contract was far from a done deal, Skoda was no longer completely shut out. In October 1905, Günther reported to the firm about the Serbian trials with Skoda's gun. The gun had performed successfully, but the Serb commission had examined additional participants, including Ehrhardt, Krupp, St. Chamond, and Schneider. Moreover, the Serbs now indicated that the contract would be awarded based on the final report of the commission. Given the delays so far, Skoda understood that in all likelihood the Serbs would not make their decision before the end of the year. On November 5 Austrian bankers reported to Vienna that the loan, involving English and Dutch banks, was as good as concluded, and a contract would be on its way to Serbia within a week. Skoda's guarded optimism evaporated in December when Austria learned of Serbia's Customs Union with Bulgaria, and Vienna demanded that Serbia dissolve the Union. To punish the Serbs, Austria pulled out of the loan. The Austro-Serb trade war erupted with the so-called Pig War on January 22, 1906, when Austria halted all imports from Serbia. In response, all the Serbian political factions rallied in united opposition to anything smacking of Austrian coercion.[40]

In March 1906, opinion among Artillery Commission members was divided. Some were in favor of Creusot, but five members wrote against

Creusot and put the choice between Krupp and Skoda. The German envoy, Erik Dürnberger, declared to the Austrians that he was almost positive that the artillery order for sixty field and twelve mountain batteries would go to Krupp. He also told the Austrians that Krupp would be willing to share part of the order with Skoda. Krupp would retain the guns, but would distribute some of the orders for gun carriages and shells. According to Dürnberger, any sharp competition between Germany and Austria had to be avoided as it would serve only the French. Dresdner Bank and Deutsche Bank were offering a good loan for 70 million francs at 4.5 percent.[41]

The Austrian ambassador in Belgrade, Baron von Wahlborn Czikann, declared to Mihailo Vuic, a Serbian member of parliament, that since Skoda had offered the lowest price, Austria would consider it an unfriendly act if Serbia did not place the order with the Austrian firm. He had heard from his sources that the government wanted to give the order to Krupp with the complete exclusion of Skoda. He told Vuic that if Serbia took such a course of action, Belgrade would find it difficult to conclude the trade agreement. That same day, April 6, the details of the commission's vote came out. In an official communiqué the Artillery Commission announced that they had put the test guns into two groups. In the top tier they placed Krupp and Creusot. Skoda and Ehrhardt were consigned to the lower tier. In the top tier, the commission's vote had split 6–4 in favor of Krupp. In explaining the vote to an outraged Loewenthal, the Austrian military attache, the Serb foreign minister Vasilije Antonic said that Skoda's gun had received only three votes, and that the experts had determined that Krupp's gun had performed better. Antonic maintained that his hands were tied, and that his government must abide by the decision of the commission. The next day, Baron von Wahlborn Czikann's dispatch explicitly spelled out that the trade agreement with Serbia would depend on the gun contract going to Skoda.[42]

On April 8 Czikann met with King Peter to discuss the gun order. The king stated that he wanted friendly relations with Austria. According to Czikann, the king "gave me his solemn word that our industry would not be eliminated from the cannon order." Furthermore, if Austria did not receive a portion, he would use his veto.[43] Czikann followed up by meeting with General Sava Gruic and Antonic. Both Serb officials echoed the king's position that Austrian industry would not be excluded from artillery orders, and they expressed hope that Skoda would be in the running for the mountain gun order.[44]

By mid-April the gun question had exploded into the public realm, and the political fallout brought down the Cabinet. Pašić's government returned to power on April 17. In diplomatic circles it was common knowledge that the Austrian military attaché had made the granting of a Commercial Treaty with Serbia conditional on the order for guns and other contracts being placed with Austrian firms. From the Austrian point of view, Serbia had systematically put Austrian tenders for all government contracts aside, even when financially the terms offered were better than those of other competitors. As affairs stood, Austrian demands on Serbia had escalated to include the orders for guns and railway materials, a certain share of the loan, and the contract for rock salt presently held by Romania. The Serbians resented this Austrian blackmail.[45]

Pašić insisted he was not anti-Austrian. He told Austrian officials that he preferred that the orders go to French and Austrian firms. When Czikann met with the president on May 6, Pašić said that the trade agreement might be brought up for the vote, and there was no time to lose. He also agreed to meet with Günther. While seeming amenable to the Austrians, Pašić insisted that all discussions be confidential, especially over the gun question.[46] The Serb interest in securing the trade agreement with Austria seemed so vital that the British ambassador, Wilfred Thesiger, reported that Pašić "appears quite ready to purchase the Commercial Treaty by an order for the same Skoda guns which a few months ago he openly condemned as worthless."[47]

The Artillery Commission presented its official report to the Skupshina on May 22. The test results for field guns yielded the following rankings: first, Krupp; second, Schneider; third, Ehrhardt; and fourth, Skoda. For the mountain gun trials, Schneider performed best followed by Krupp, Skoda, and then Ehrhardt. On June 11 the Austrians were told the contract would go to Schneider. In secret balloting taken on the question of general excellence, the majority of the commission had voted for Krupp. However, a more detailed vote extending to the different parts of the guns and carriages had favored Schneider. According to the classification system, Schneider's gun had earned the greatest number of "very good" marks.[48]

In the summer of 1906 Austro-Serbian relations over the gun question reached a nadir. The Austrians steadfastly refused to let the matter drop. Vienna continued to insist that the Serbs give artillery orders to Skoda at least for the mountain guns. The press in both countries published heated articles accusing the other side of lies, deceptions, and bad faith. To undermine Austria, Pašić gave the Skupshina a bluebook that disclosed the background of the trade conflict, including the confidential discussions he had held with the Austrians. In selecting the documents, Pašić sought to place all the blame on Vienna. In retaliation, the Austrian Press Bureau published its own documents.

Although at times Pašić and other Serb officials had made conciliatory promises to the Austrians, they had refused to give any written assurances about government contracts. Meanwhile, the Serbs conducted experiments at their arsenal to see whether they could modify guns on their own. The Serbs hoped that these experiments could provide an excuse to downplay the need for ordering new guns, thereby taking the gun question off the table and clearing the way to reopen the commercial negotiations with Austria.[49]

Sitting on the sidelines, the British again found themselves shut out of Balkan sales. The British representative, J. B. Whitehead, took up the matter with Pašić in late August. Whitehead mentioned that the Armstrong representative had complained that the Serbian commission had not visited Armstrong or invited any British firms to compete. With characteristic standoffishness, Whitehead reported, "I did not of course press the matter in any way, but merely expressed surprise that so important a firm as Armstrong's should have been omitted from the investigations of the Commission." Pašić's explanation stressed financing. The gun question was necessarily connected with the loan question. As far as he was concerned, this was another reason against ordering the guns in Austria or attempting to raise a loan in that country. The Serbian bondholders were represented in Serbia by two commissioners: one for the French and the second for Austrians and Germans through Deutsche Bank. The Austrian government made it known that if a new loan originated in Austria, they would demand a third commissioner to take special charge of Austrian interests. The Serbs feared that this commissioner would attempt to exert political influence. Pašić concluded by suggesting that Armstrong could receive an order for machine guns in the future.[50]

Finally, Serbia signed the contract with Schneider in November 1906. The total order amounted to fifty-six batteries at four guns apiece with a distribution of forty-five field batteries, nine mountain batteries, and two batteries of horse artillery. The banks furnishing the loan for 95 million francs at 4.5 percent interest consisted of Banque Ottomane, Société Financière d'Orient, and Banque Franco Suisse.[51] The remaining step involved legislative approval. In bringing the loan and gun order before the Skupshina, the Pašić government had to confront an interpellation brought by the opposition. Between November 26 and December 5 the Skupshina's session was taken up by the question. In answering the question, Pašić had all the Artillery Commission reports from 1903 to 1906 read to the House, which took the entire first day. He declared that the choice of what company should make the guns rested with the competent specialists of the Artillery Commission. In closing, Pašić stated, "The Serbian government has endeavored to select the best weapon. You have seen to what extent foreign influences were exerted to prevent the armament of Serbia. If we were to continue to do nothing but make comparative trials we should never get any guns."[52] As expected, the energetic attacks of the opposition parties were not successful. Ultimately, the government's choice of orders passed 89–62. With the path now cleared, the body voted approval for 59 million francs in military expenditure inclusive of a loan of 46 million francs (£1.84 million) to acquire more than 200 quick-fire guns and 70,000 rifles.[53]

Having asserted Serbian independence from Austria in placing the artillery order, Pašić was willing to placate Austria by granting the rifle contract to Austrian firms. Serbia took bids in April 1907. Steyr, DWMF, and a Hungarian rifle firm tendered for the order. As mentioned previously, DWMF and Steyr had agreed to share Balkan orders, with the German firm ceding Serbia to Steyr. Consequently, the competition was a façade. Steyr and DWMF submitted the lowest, identical bids of 79 francs per rifle and 75 francs per carbine, most likely with the expectation that Serbia would favor the Austrian firm. For munitions Hirtenberger proposed the lowest price, underbidding fellow Austrian firm Roth, the Hungarian firm Weis, and the Belgian Herstal. British firms Nobel and Birmingham Metal and Munitions Company applied, but initially failed to comply with various conditions. The English bid of 109 francs per thousand was lower than Hirtenberger's bid of 109.85 francs. Law required that the contract must be assigned not to the lowest tender, but to the most favorable, and the English lost the bid because of their difficulties. The assignment of the contracts was repeatedly discussed in the Council of Ministers. By the end of April the cartridge orders had been assigned to Hirtenberger and Weis. The rifle order remained unassigned, but Pašić assured the Austrians that the war minister favored

Steyr. The government had not finalized the contracts so that Pašić could use them as a factor in the commercial negotiations with Austria. Ultimately, Steyr received the contract in late July.[54]

Given the relatively poor performance of the Skoda gun in the trials, why had Austria pushed the Serbs so hard on behalf of Skoda? Simple economic interests could have been satisfied in other ways, such as participation in the loans or compensation in railroad and rifle orders. Constantin Dumba, in his memoirs, offered one possible explanation. The former Austrian ambassador recounted that Skoda was experiencing a financial crisis at the time. The firm had borrowed 4 million kronen from the Austrian Kreditanstalt, and no new credit would be extended. Also, at a ball in Vienna Kaiser Franz Joseph, the archduke, and Count Agenor Goluchowski each had made a point of impressing upon Dumba the importance of securing the artillery order for Skoda, and the archduke may have held personal investments in the firm.[55]

An examination of the financial records in the Skoda archives calls for a reevaluation of Dumba's assessment. The records reveal that Skoda did experience a bad year (1903–1904). In the three previous years the firm had earned roughly 3 million kronen annually. For 1902–1903 the figure was 3.039 million kronen, out of which war materials accounted for 832,000 kronen. Significantly, military production had become the most profitable branch of Skoda's business by 1903. Over the course of 1903 Skoda's turnover amounted to 12.607 million kronen, and military production contributed 10.817 million kronen to that figure. In 1903–1904 the company closed with a loss in the amount of 546,000 kronen, but the military branch showed a profit of 124,839 kronen. The problem lay in the dead weight of unsold inventory. The finished and partially worked items together had an estimated value of 6.887 million kronen. As the report for 1903–1904 noted, the *waffenfabrik* (gun factory) had not been provided with orders for a long time, and the backlog of inventory had increased by the end of the fiscal year. Obviously, if the Serbian contract had come through as promised by King Alexander in 1903, Skoda would have found itself in better shape. However, the firm did see improvement the next year. Skoda obtained major orders from the Austrian government by November 1904. By October 1905 the company had experienced "fairly satisfying work during the first thee quarters of the business year," and had shown a profit of 1 million kronen, "the greatest part deriving from the waffenfabrik." Thus, on the eve of the Austro-Serb trade war, Skoda was no longer in financial crisis.[56]

Strategic diplomatic concerns presented a more compelling motive for Austrian officials than saving Skoda. With Russia temporarily removed from the Balkan scene due to the war with Japan, Austrian leaders believed they had a free hand in the region. Serbia and Bulgaria both feared the uninhibited exercise of Vienna's dominance over their respective economic and military sovereignty. In reaction to an unfettered Austria, a Serbian-Bulgarian entente had already begun to coalesce in early 1904.[57] Along with aligning themselves in economic relations through the Customs Union, the two Slavic states considered the possibility for joint military action in the future. Once Bulgaria had chosen Schneider guns, Serbia had incentive to follow suit so that the two armies would have identical systems and interchangeable ammunition.[58] The challenge posed by a united Serb-Bulgarian front unsettled Vienna. For Belgrade the gun order represented the assertion of Serbian national independence and defiance of any Austrian diktat. In the words of Pašić during the interpellation, "The demand of Austria-Hungary that Serbia should give written engagement to come to no decision in the armament question before the Commercial Treaty had been concluded, could not be conceded by the Serbian government because such a demand was incompatible with the dignity of Serbia."[59] Conversely, for Vienna the gun order marked the transformation of Serbia from a friendly neighbor into a hostile one.

The elimination of Skoda set up a replay of the Krupp-Schneider duel. Vickers and Maxim tried to enter the contest in October 1906, but by then negotiations with Schneider had progressed so far that the British firm had no chance. Once again, French financial backing proved decisive. As the British accurately summed it up, "The reasons which induced the Serbian Government to select the French firm in preference to Messrs. Krupp, or any other, were entirely of a financial nature, the loan by means of which the guns were to be paid for being obtainable under better conditions in France than elsewhere."[60]

In October 1908 Austria began interdicting Serbian deliveries of war materials because of the Bosnian Annexation Crisis. The Austrian action disrupted Serb domestic production, forced Belgrade to find substitutes, and caused delay in rifle conversion at the Serbian arsenal by halting the supply of the requisite new barrels from Steyr. From the 30,000 Mauser magazine carbines ordered from Steyr in April 1907, about 20,000 rifles and no carbines had arrived in Serbia by October 1908. Of 40 million rounds of small arms ammunition ordered in Austria in April 1907, not more than 20 million had been received in Serbia by October 1908. In November 1908 Serbia placed orders with the German branch of Vickers Maxim for fifty Maxim-Mauser machine guns (Mitrailleuses) to be delivered in March 1909, and the Serbs considered buying 150 more. By December 1908 Serbia had given orders for 50 million rounds, divided evenly between French and Belgian firms, and expected the first consignments in February 1909. Also in December 1908 Serbia placed an order for 50 million Mauser cartridges in England with Kynoch Ltd. on a ten-month contract with goods to be delivered through Salonika. The Serb government also ordered 30,000 Mannlicher rifles from Budapest Waffenfabrik, and tried to buy 80,000 Romanian rifles through a Berlin firm, but Romanian prime minister Dimitrie Sturdza forbade the transit. Serb agents also tried to buy war materials in Berlin and other German cities. In January 1909 the French arms dealer Laurent wanted to buy 55,000 Russian Mosin rifles and 50 million cartridges from the Japanese government, and to sell them to Serbia. By February talk in Serbia favored 33 million francs for a new armaments credit, not including artillery.[61]

Aware of Serbia's dependence on the port at Salonika for its arms imports, Vienna tightened the noose around Serbia while other powers unsuccessfully tried to aid the Serbs. Creusot was delivering war materials to Serbia through Salonika in October 1908.[62] Janos Pallavicini, the Austrian ambassador in Constantinople, asked Grand Vizier Hilmi Pasha, the Ottoman Minister of Transport and the governor of Salonika, to stop artillery shipments to Serbia. Officials replied that they could only halt the shipment with an order from the central government.[63] In the words of Pallavicini, "France because of its business interests, England because of its political flirtation with Serbia, and also Russia as guardian of Serbia, requested the transit of war materials by the Porte."[64] Russia sent twenty machine guns as a gift to Serbia in February. By March, the Russian government pressed Bulgaria to allow transit of Serb war materials. The Bulgarian war minister reported that no request had come from Serbia itself, although he declared to the Austrians that he would oppose such a request. The Turks had by now decided not to antagonize Austria, and therefore Grand Vizier Hilmi Pasha forbade the transit of Serb war materials through Ottoman lands. This turn of events caused the Serbs to look to Piraeus in Greece as an alternate route.[65]

The Bosnian crisis ended, but it brought a significant change in Serbia's arms trade. Austrian suppliers lost out as Belgrade gave new business to English firms for ammunition. The Serb government had become preoccupied over the question of delivery of war materials by 1911. Since the annexation crisis, when Austria stopped arms and ammunition imports through its territory, Serbia had obtained supplies through Turkey via Salonika. Serbia now preferred to order munitions from England or France, in the belief that those countries would, in case of such impediment, pressure the Porte to allow delivery. Serbia had also begun reciprocating the favors received from Turkey by allowing war supplies from Germany to pass through Serb territory during the Italian war. When in 1912 the government called for tenders for 75 million cartridges, an Austrian firm offered the lowest price but the Serbian government demurred buying lest the Austrian government should interfere with delivery. The Serbs applied to Lloyds Bank for help in paying for 75 million cartridges ordered from Kynochs.[66]

Financial support continued to play a role in war orders. In November 1909 Serbia reached agreement in Paris for a loan of 150 million francs over fifty years. Under the terms the Serb government had to use three-quarters of this loan for war materials and railroad materials to be made in France, and one-quarter in Germany. Without observing these terms, the loan would not float on the French and German markets. The French and German governments accepted collaboration by their respective financial groups. Serbia promised the French 27 million francs in war orders, but the Germans wanted a portion of the artillery order as a protest against the preference previously shown Creusot. For political reasons, Serbia did not want to depend on German armaments. However, the Serbs used the competition to apply pressure on Creusot to reduce its prices. In the end Serbia placed a small arms order in Germany worth 3.8 million francs, and Belgrade increased the German share to 10.7 million francs compared to Schneider's 27.5 million francs. Subsequently, Serbia placed an order for 60,000 Mausers with 1,500 rounds each in Germany.[67]

Having bested Krupp first in Bulgaria, and then in Serbia, Schneider now set its sights on Greece. The Germans found themselves on the defensive as their long-held dominance in the artillery business evaporated. Meanwhile, the Austrians continued to hold steady in rifles and cartridges. In October 1904 the Greek government stated that artillery trials would be postponed due to financial difficulties. Captain Brémond d'Ars, the French military attaché in Athens, learned that the Council of Ministers had decided unanimously to order field guns from Krupp without conducting tests. The government proposed the purchase of Krupp quickfire guns in mid-May 1905, but no contracts were completed. By the summer of 1906 the Greeks contemplated upgrading their field artillery by purchasing thirty-two batteries of quick-firing 7.5cm guns. They invited several big firms, including Armstrong, to bid.[68]

In 1897 Greek artillery orders went to Krupp in part because of a familial alliance between the ruling houses of the two countries. But since the Greek prince had married a Frenchwoman, the Greek government decided to hold a competition in 1907.[69] The Greeks conducted the field tests between April 5 and July 7, 1907, and the participants included Armstrong, Ehrhardt, Krupp, and Schneider. What occurred next was extraordinary. In the words of the British report:

> As it proceeded it became evident that the English and German guns were outclassed by the French. To "save their face" the representatives of the two German firms announced that they withdrew from the competition because they found that they did not receive fair play at the hands of the managing Committee of officers, the members of which, not excluding the President, His Royal Highness Prince Nicholas, they hinted very broadly, had been bought by the French. Their protests were supported by the German Minister in language so strong as to give great offence not only to the Greek Government and army, but to the King himself. On the other hand, the German Emperor has signified here his displeasure at the transfer of Greek custom from Essen to Creusot. It will be some time before this incident is forgotten in Greece.[70]

Count Emmerich d'Arco-Valley, head of the German mission in Athens, succeeded in making himself extremely disliked by the strength of his language about the field gun trials.[71]

Given Krupp's public assault on the integrity of the process in Greece, Prince Nicholas believed that Greece was obliged to publish the official report because of the "diplomatic means that Krupp had threatened to employ."[72] Subsequently, the 200-page report was published. Based on the test results the majority of the commission concluded that only two models were really in the running: Krupp and Schneider. The two guns were equally sound from the point of view of mobility, but from the standpoint of firing, the Schneider gun demonstrated greater stability in all circumstances. Also, the sighting system was easier and more precise on the Schneider gun. Thus, the decision favored Creusot by a vote of 6–3. The six favorable votes were Prince Nicholas (president), Colonel Clomenis, and four superior officers. The three opposed were lower officers, one of whom was a representative of Krupp in Greece. Krupp, foreseeing the writing on the wall, had withdrawn its material four days before the end of the trials and complained formally to the Greek War Ministry after. The contract went to Schneider on November 26, 1907, for 36 batteries (144 guns) of 75mm quick-fire field guns and 6 batteries of mountain guns, with 1,500 rounds per gun, for a total value of 12 million francs. The Greeks also granted Armstrong an order for Howitzer shells.[73]

Public opinion in Greece turned against the Germans and their high-handed behavior. In the newspaper *Acropolis* on July 13 an article began, "It is time that our good friends of '97, the Germans, understood that Greeks are not Turks. The Kaiser can impose his will on the Sultan and say I want you to accept Krupp cannons. It is done that way in Constantinople, but not in Greece."[74]

In September 1909 the Greek government proposed to increase its artillery ammunition. Armstrong sought support from the Foreign Office in negotiating with the Greek government. During the gun trials in 1907, Armstrong was told that it would receive the order for 21,000 high-explosive shells because of its product's superiority. As of October 1909 this contract had been discussed for two years but not submitted. On October 27, 1908, Prime Minister George Theotokes said he would not order more than 10,000 rounds, but two days later he increased it to 12,000 rounds. During the spring of 1909 Theotokes promised to give the order at once if Armstrong reduced its price. Armstrong did reduce its price, and still received no order. Then in October the firm heard that the order probably would go to Krupp because Essen had submitted an amended tender at a lower price than Armstrong.[75] Armstrong's chairman explained to the Foreign Office that "a similar attempt was made by [the] Ehrhardt firm to obtain the order for shrapnel shell just before it was placed with Messrs Schneider, but owing to the help this latter firm received from the French legation at Athens the attempt failed. We shall be very grateful if the Foreign Office can bring any diplomatic pressure to bear on the Greek Government. Otherwise we fear that we shall be tricked by illusory promises and still lose an order that has been allotted us after unusually severe competition."[76] The Foreign Office assured Armstrong that its ambassador in Athens, Sir Henry Elliott, had repeatedly spoken about the shell order to successive Greek prime ministers.[77]

Armstrong received additional support from the French. When a French artillery officer, Major Lacombe, arrived in Athens in October 1909 to train the Greeks in the use of their new Schneider artillery, the British ambassador and military attaché met with him and enlisted the aid of the French officer. Lacombe informed the British that the Greeks had not even ordered common shell for practice fire.[78] Despite the proven superiority of the Armstrong high-explosive shell, which Greek tests had rated three times superior to the French and four to five times better than German shell, the British firm continued to wait for the Greek contract. The British military attaché Colonel William Surtees complained about Armstrong's unfair treatment. "One of the clauses in the original competition states that prices would not outweigh against quality, and that this theory has just been emphasized by the Greek government, which has given a contract to Krupp for naval shells at 25 percent higher price than Ehrhardt's tender—it seems not unreasonable to suggest that the Greek authorities be insisted to extend the same fair play to messr Armstrong—the vast superiority of whose high explosive shell over all others—being beyond discussion."[79]

At the start of 1905 the Greek decision for a new rifle had come down to a choice between Mauser or Mannlicher. The German and Austrian embassies in Athens each contested to bring the contract home. The Germans found their main champion in the person of Crown Prince Nicholas and to a lesser extent General Chief of Staff Constantine Sapundzakis. The Austrians had as their advocate War Minister Mavromikalis. The decision rested with the Defense Commission, which chose the Mannlicher and Austrian cartridges.[80]

In stark contrast to the other Christian Balkan states, Romania alone remained firmly in the Krupp camp. Krupp not only continued to dominate the Romanian artillery market, but the German firm also displaced its Austrian rival, Skoda, in management of the Romanian domestic arsenal. The Austrians did retain their preeminence in supplying Romania with small arms and cartridges. Steyr served as the main rifle supplier. In 1903 Bucharest placed an order with Steyr for 30,000 to 35,000 Mannlichers, on top of the 140,000 already in hand.[81] For the ammunition Romania employed the firm Keller and Company of Hirtenberg to supply 30,000 cartridge cases for Mannlicher rifles (93 or 94 francs per 1,000).[82] The deeply held preference for Krupp by the king and his cabinet allowed for German success despite the acknowledged technical superiority of Schneider guns, as attested by army officers. Thus, politics won out over professional considerations in Romanian armaments.

In February 1903 a battery of four quick-fire Krupp field guns of the latest model arrived for tests in view of possible adoption by Romania. The guns were almost identical to French guns, but the highest artillery officers believed French guns were still the best. In explaining why the Romanian government did not seek to acquire the French guns, British ambassador J. G. Kennedy heard from his informant that, in case of a European war, Romania would join Germany and Austria against France and Russia, and consequently it might be difficult to obtain further supplies of the same gun and ammunition. Furthermore, Krupp accepted payment by means of a loan, namely by yearly payment of interest and sinking fund. The British ambassador's analysis placed greater emphasis on the personal preferences of the monarch. "The real reason," Kennedy conveyed to London, "for the preference given to Krupp is due to the strong and natural sympathies of the King of Romania for everything German, which make it impossible for His Majesty to apply elsewhere than to Germany, or perhaps to Austria, for war materials."[83]

Krupp faced no competition. In June 1903 Romania conducted trials with the Krupp guns in anticipation of a probable order for 300 pieces. Paying for the order required a special credit of £1.112 million. In December 1903, Dimitrie Sturdza, the Romanian president of the Council of Ministers, responded to an interpellation of General Iacob Lahovary regarding the proposed orders in Germany for Romanian artillery. Sturdza was accused by the Senate of conducting negotiations with Krupp to the exclusion of all other firms, especially the French. Sturdza did not defend himself well, saying that the French were very conservative and that he had conducted only preliminary negotiations with them since the Krupp tests were not yet completed. Ultimately, the Romanian Palace

of Deputies voted a credit of 28 million francs for 300 new quick-fire
field artillery to be supplied from Krupp, delivered over two and a half
years at 4.5 percent interest. On top of this, an additional order for
thirty-six guns was placed with Krupp. In April 1905 Romania took a
loan of 40 million francs, 21 million of which was directed for debt service, and the remaining 19 million paid to Krupp for artillery orders.[84]

The question arose of what to do with the old cannon. Sturdza discussed the possibility of cannon conversion with Austrian field marshal
Krepachek, the main inspector of artillery, who proposed an agreement
with the Austrian government. After learning about the possibility to
modify an old piece to increase the firing rate from one shot to six shots
per minute for only 107,000 francs, Sturdza personally petitioned in
Berlin before the Prussian War Ministry to expand the order for two machines in the government arsenal at Spandau under a commission of
German and Romanian officers for the next two years. Austria had offered access to its factories for five years. The Russians considered the
matter an example of German influence over Austrian.[85]

Competition for field howitzers in 1905 involved Krupp, Skoda,
Ehrhardt, and Schneider. In May Ehrhardt tried to enlist the support of
the German Foreign Office for its howitzers rather than Krupp's. Ehrhardt
suggested that they were the best of German industry in the struggle
against the French. Later Ehrhardt complained to Wilhelmstrasse that
they had learned from their Bucharest representative that the Romanian
government believed that the German government recommended Krupp
over Ehrhardt. Whether this was a misunderstanding on the part of the
Romanian government or the opinion of the German ambassador, it was
contrary to the known position of the kaiser that Krupp or Ehrhardt
would be equally good. Finance Minister Take Ionescu told the German
military attaché, Kurt Von Hammerstein, that the Council of Ministers
had a strong preference for Krupp, from whom they had ordered 12cm
howitzers in 1900. The war minister told the German ambassador that
Krupp products were highly esteemed.[86]

Ionescu also indicated that the shell question had a military-political
meaning since Romania had for years depended entirely on other lands
for these supplies. In particular, Romania relied on Austrian munitions,
and for its arsenal equipment Romania had turned to Skoda. Now along
with the guns and naval deliveries, Ionescu suggested that Krupp could
extend its reach to obtain shell orders. The finance minister advised that

Essen must reduce its price 20 percent. Krupp, which under Sturdza practically had a monopoly, had attempted to raise its prices, and so Romania had looked to Austria.[87] War Minister Mano said that the order of an additional 100 Krupp field guns would not be pushed at the moment, but there would still be time in fall. He also said the Austrian 15cm howitzer was too heavy for the field army. The Romanian Artillery Commission seemed pleased with the new French 10cm field howitzer, model 1899. Mano wanted the German howitzers, and he suggested the Austrians would be compensated through arsenal and explosives orders. Mano told Von Hammerstein that the Austrian and French ambassadors sought to force out Germany from the local cannon market. The Romanian military attachés in Paris and Vienna were considered Germanophobes or even pro-French. The one in Vienna was the son of the Franco-Romanian Fraternal League president. Nevertheless, Romania had ordered from Krupp for forty years. Von Hammerstein observed, "That the Romanian War Minister at the time so decidedly favored German material without undertaking comparative tests and that his position was so contrary to the view of Lahovary, is for the German arms market a great victory."[88] In July 1907, while War Minister Mano was abroad, Lahovary tried to use his position as interim minister to give an order for machine guns to France and take it away from DWMF. In Von Hammerstein's assessment, "Lahovary remains a Frenchman, who with all his energy works against every increase of German influence in military matters, and he places all possible difficulties in the way of the introduction of German war materials into Romania."[89]

In terms of domestic production Romania switched from Austrian to German contracts. The Romanian Ordnance Department, because of difficulties in manufacturing the delicate machinery for artillery shell production, had hired experts from Skoda to instruct the local workmen. In the beginning of 1907 Romania worked with Skoda to equip the Bucharest Arsenal to produce quick-fire shells. But the Skoda work was not deemed satisfactory, and in the fall of 1907 the Romanian government cancelled the contract with Skoda and ordered new shells from Krupp. According to the Russian military attaché, the real reasons for the change from Skoda to Krupp were the intrigues and bribery of Krupp agents, who themselves wanted to take over the Bucharest Arsenal.[90] The project with Skoda failed because of quarrels between the Austrians and Romanians. General Alexandru Averesco's visit to Berlin in 1908 concerned the Krupp proposal to establish an arsenal in Romania. Because the Romanian army had recently been entirely rearmed with quick-firing guns by Krupp, and Krupp had also supplied portions of armaments of the forts, it seemed probable that Krupp would secure this additional advantage. "It would appear," mused the British ambassador in Bucharest, "as if Romania were minded to have her policy shaped by Austria and her projectiles by Germany."[91]

The Krupp proposal in regard to the arsenal at Bucharest had the strong support of King Charles. From the king's perspective not only would establishing Krupp in Romania solve the problem of the Romanian's inability to turn out the best material, but it would also put an end to the intolerable delays and inconveniences when trying to import war material from Essen through Austrian-Hungarian territory. "It would hardly be believed, His Majesty said, that it took four months to get a gun across Austria-Hungary, chiefly owing to the difficulties encountered at the latter's frontier. All this would have been avoided and an excellent technical education would have been provided for Romanian workmen if the suggested arrangement with Krupp had been carried through. It was nothing but 'rank chauvinism' the king said, but as soon as he had discovered that the idea was unpopular he had told General Averesco that it would be better to drop it."[92] German enterprise did succeed in wresting the Dudesti state powder works from Austrian hands. In June 1910 the Rottweil Pulverfabrik of Cologne took over from the Austrian firm Blumenauer Fabrik.[93]

Even as Romanian officials paid lip service to giving fair evaluation to other suppliers, Krupp's dominance in Romania continued. In May 1908 the Romanian government began discussions with Creusot about fifty batteries (four guns each) of 75mm and several howitzers (10.5 and 12.5cm). The amount of 15 million francs was voted for this purpose.[94] In February 1911 the Romanian government again placed orders with Krupp for field guns. Although the commission also was studying a French howitzer, "this was merely to satisfy French susceptibilities; the order for these also would eventually be placed in Germany."[95] At the first reception of the diplomatic corps by the minister of foreign affairs, the British diplomat expressed that if large orders were to be given for war material, British firms should also be given an opportunity of tendering. "Monsieur Maioresco said that of course he knew that there were several first class British firms who supplied guns etc. and that he would certainly speak to the Minister of War on the subject. I fear, however, that the new orders are likely to be placed in Germany as heretofore."[96] Finally, in February 1912, Romanian war minister Nicholas Filipescu signed the contract with Krupp for 10.5cm howitzers, fifteen batteries (four guns each), after three years of trials.[97] In March 1912, Romania ordered eight heavy field howitzers from Schneider (1.45 million francs) and sixteen mountain guns (967,950 francs). In comparison with the large German orders amounting to almost 20 million francs, these French orders were a pittance. However, pro-French circles had begun an intensive campaign around the Schneider artillery order to discredit German industry.[98]

German dominance still prevailed in orders for the Ottoman army, but the French avidly pursued Turkish artillery orders. The Ottoman Bank and the Ottoman government had concluded a new Turkish loan for TL2.5 million (62.5 million francs) for arms on September 17, 1903. Deutsche Bank participated for 25 percent of the loan. As part of the agreement German firms were to receive contracts: Krupp for thirty-two batteries worth 18 million francs and Mauser for rifles worth 14.5 million francs. The French demanded "compensation," and they blocked the loan. The Turks sought to escape the French conditions while the French government opposed collaboration between French and German banks. Nevertheless, negotiations continued between the Ottoman Bank and Deutsche Bank in Constantinople in November 1904. The Ottoman government officially invited proposals for a huge artillery order in December 1904. The Turks wanted to purchase 101 batteries of field artillery (6 guns per battery) with 500 shells per gun, and 12 caissons. In addition, the palace wanted 6 batteries of horse artillery, 23 mountain gun batteries, 6 batteries of howitzers 15cm, and 6 batteries of siege guns.[99]

The Ottoman Bank was willing to allow Krupp the artillery order if French industry obtained naval orders. In January 1905 Deutsche Bank and the Ottoman Bank jointly prepared to issue 200 million francs in a new loan, but German ambassador Baron Marschall von Biberstein could not accept an order of cannons given to Schneider because such a contract would violate the German monopoly in Turkish artillery. The Germans, through von Biberstein in Constantinople, offered the French the railroad contracts as long as Germany retained the artillery orders.[100]

Ultimately, the Turks preferred to conclude loans with each group separately. The Ottomans signed with the Ottoman Bank for a French loan worth 60 million francs, out of which Schneider claimed approximately 17 million francs for naval orders. The Turks also concluded a German loan for 60 million francs, from which Krupp received 46 million francs in contracts.[101] Thus, in April 1905 the Germans took two-thirds of the Turkish war orders, including the field artillery order for TL1,974,298, and the French share amounted to TL714,000.[102]

In effect Ottoman loans and conversions in the years 1881–1903 totaled 385 million francs, three-fourths of which France had furnished. These funds had served to buy German war materials. The affair of Turkish loans was in reality a rivalry between Krupp and Schneider through their governments and, for better or worse, by Deutsche Bank and Ottoman Bank. This attitude directly threatened Krupp's position and that of German industry in Turkey. Deutsche Bank preferred to work with its French colleagues in the Ottoman Bank to assure its participation, but for questions of prestige and political and economic interest, the German government thought it could not let a French preponderance develop in Turkey. However, neither Delcassé nor Wilhelmstrasse desired a rupture, and they did not oppose the banks' agreement in April 1905.[103]

In holding off the French, the Germans generally and Krupp specifically had achieved their greatest Turkish sales yet in 1905. The Ottomans ordered ninety-one batteries of Krupp artillery. Because the order was so large, once again the Deutsche Bank stepped in and contracted the loan. The Ottoman government was unable to pay arrears amounting to almost TL1 million on its former contracts for Krupp guns and Mauser rifles, so it was decided to pay off the old debt and at the same time arrange a new contract from Deutsche Bank. This time the loan amounted to TL2.64 million, of which TL2,098,800 (about 79.5 percent) was actually received. Additional customs duties for military equipment and the 6 percent additional revenues of the Public Debt Administration served as security for the loan. This order was over twice the size of the 1889 order.[104]

In hopes of breaking the German hold on Ottoman sales, British and French artillery exporters entered into a cooperative agreement not unlike the Austro-German rifle cartel. In March 1907, Vickers and Armstrong concluded an agreement with Schneider pertaining to business in Turkey that would remain in effect until January 1, 1910. The arrangement was to avoid "unnecessary and useless competition" and to improve chances for contracts, and the terms called for an equal division of orders between the French and British groups. The British Foreign Office subsequently instructed the embassy in Constantinople to work with the French ambassador as much as possible to promote the "united interests of their respective countries" in obtaining artillery orders. Meanwhile, Schneider officially informed the French government of the existence of the companies' agreement as well.[105]

The overthrow of Abdulhamid by the Young Turk revolution established a parliamentary system and brought changes to the procurement process. No longer would the will or whim of the sultan determine the armaments contracts. Parliament had to approve the budget and could call the War and Naval Ministries to account. The Young Turk revolution restored the constitution on July 24, 1908, and the Parliament held its first session on December 17, 1908. On April 26, 1909, the Parliament unanimously deposed Sultan Abdulhamid II, leaving the Committee of Union and Progress in power.[106]

At first the Turkish preference for German suppliers continued under the Young Turks. In early November 1908 the Turks were looking to order 300 million cartridges for Mauser rifles, and the War Office invited competition from English, German, French, and Austrian firms. British firms Nobel and Birmingham Metals and Ammunition came to Constantinople and quoted a price of £5.10 per thousand. The Austro-Hungarian firms Keller (Hirtenberg), Roth (Vienna), and Weiss (Budapest) all competed. Hirtenberger submitted the lowest bid. DWMF Karlsruhe bid £5, and the Turks ordered 250 million from the Germans. In December the Turks concluded their contract with Karlsruhe for the 250 million cartridges at TL1.385 million plus a commission of 15 percent (TL207,750). The terms allowed for an additional 25 million apiece to be subcontracted to Bolte (Magdeburg) and Ehrhardt (Düsseldorf). Delivery commenced in January 1909 and finished in 1909.[107] Subsequently the British learned that "the German agent was informed of the English price on the 3rd November, and that he at once put in a lower tender, which was accepted without any reference to the British representatives, who therefore had no chance of fair competition."[108] Henry Lowther registered British frustration with General Mahmoud Mouktar Pasha, commander of the Turkish First Army Corps. The general explained that when the order was put up for tender, the military authorities did not know that the pointed steel bullet, which was insisted upon in the specifications, was a monopoly of

Karlsruhe Small Arms, and consequently other firms could not compete.[109] Colonel Surtees, the British military attaché, drew the conclusion that, "notwithstanding the loud expressions of friendship and gratitude to England which have been heard since the declaration of the Constitution, it would seem that British industry stands no better chance against German competition than it did under the old regime."[110] If the British failure was not much of a surprise, the Turkish rejection of the Austrian firm Hirtenberger must have come as a shock to Vienna. Besides having submitted the lowest tender, Hirtenberg had been overseeing Ottoman cartridge production since 1904 when the firm established cartridge production facilities at Zeytinburnu for TL1 million, and Hirtenberger had provided 50 million cartridges.[111]

As had been the case with rifle cartridges, in the contracts for field and mountain gun shells British firms found themselves used by the Turks to improve German bids. Armstrong's price was originally 10 percent cheaper than Krupp and Ehrhardt. Yet every opportunity was given to the German firms so that in the final adjudication Armstrong's bedrock price was publicly announced and the German competitors were permitted to tender against this price. Ehrhardt offered to supply the ammunition for TL9,000 less than the amount quoted by the Elswick Ordnance Works. Armstrong had been prepared to make every possible sacrifice to obtain this order, and the head of the Ottoman Artillery Department publicly thanked their agent for having saved his government some TL50,000 through the reduction made by the German firm. The British military attaché observed that "the fair dealing of Elswick does not seem to have been efficacious in changing the deep-rooted prejudice in favor of Germany."[112] The Turks in all likelihood never had any real interest in giving the orders to a British firm. From a Turkish point of view, their practices had reduced prices by TL300,000 in the case of Mauser cartridges and by TL120,000 for artillery ammunition. Colonel Surtees summed up the situation well: "Messrs. Armstrong and Nobel can, however, hardly be expected to appreciate this policy, or to continue the purely philanthropic work of assisting Turkey to buy cheaply from Germany, nor, I venture to submit, can our financiers be expected to assist a country which spends the money raised in England and France on German manufactures."[113]

The shell order went to Ehrhardt, but not before another round of competition. On November 23, Ehrhardt offered to accept TL316,000, or TL9000 less than Armstrong. Krupp then offered a 5 percent reduction on Ehrhardt's price. As the competition continued, the Turks held final offers through sealed tender on November 28. The firms bid as follows: Armstrong, TL319,000; Krupp, 300,000; Schneider, 299,000; Ehrhardt, 277,500. At last Ehrhardt won the contract, in the process beating Krupp and the non-German firms.[114]

Incidents connected with placing the order worth £250,000 for field artillery shells with Ehrhardt revealed much about Krupp's entrenched position, and the difficulties in flanking Essen. The German ambassador in Constantinople had opposed Ehrhardt, and consequently Ehrhardt's representative had to trick and outwit the German military attaché, knowing full well that he would inform Krupp's agent of everything Ehrhardt did. Ehrhardt's agent related to Lowther that inasmuch as his firm sympathized with the Constitutional Party in Germany, he received the goodwill of the members of the Young Turk Party. The Young Turks seemed delighted to place orders independently, without their hands tied by the sultan's orders to give the business to Krupp. From Ehrhardt's perspective, the endeavors of the kaiser during the past ten years had been mainly devoted to helping Krupp. In Turkey the kaiser repeatedly used his personal influence with the sultan for Krupp's benefit.[115]

Where Ehrhardt had maneuvered around Krupp, the Austrians tried to forge an understanding. The Austro-Orientalische Handels-Aktiengesellschaft appealed to Baron Wladimir Giesl for his energy and influence to help Austrian industry obtain Turkish orders. The banker had talked with a representative from Skoda regarding the inability to secure a Turkish order:

> For this establishment has for years not scored a success in Turkey, everything flows namely to Krupp, and Skoda goes away empty . . . Please remind His Excellency that we in Austria have a great gunworks and it must be kept busy. It is for the Monarchy formally a natural question that Austrian guns should also be exported, and that Krupp alone should not swallow everything. Would it be possible that one could forge an agreement with Krupp so that one would mutually agree not to compete? Today the frontrunner Krupp says that for Skoda it is impossible to receive anything in Turkey.[116]

It seemed as if nothing had changed since the Young Turk revolution. As Armstrong's board member J. M. Falkner lamented to Edward Grey, "We had hoped that something at any rate of the old baksheesh regime would have passed away. But it seems as if this were not so, and as if the German grip on the country were as strong as ever."[117] But change was coming. Mouktar Pasha told Lowther confidentially that Krupp's agent had been called on by the Turkish government to give a list of names of all those who under the old regime had taken bribes in return for orders, and implied that unless this list was forthcoming no further orders would be given to Krupp. He added that he did not think the late minister of war or grand master of artillery were guilty in this respect, but that all the money had gone to the palace officials.[118] Then in July 1909 Chief of General Staff Izzet Pasha declared that the upcoming purchase of thirty-six quick-fire mountain guns should not be considered for Krupp alone, but for Austrian establishments as well.[119]

The first break for the Austrians occurred late in 1909. In October, the Turks prepared to order 442 munitions wagons for 7.5cm quick-fire guns, 66 quick-fire 7.5cm field guns, and 36 quick-fire 7.5cm mountain guns. The competitors for the wagons included Böhler, Ehrhardt, Armstrong, Danubius (Budapest), and Raaber Waggonfabrik in Györ. Böhler bid at TL240, and Krupp and Skoda each bid TL210 per wagon. The commission considered the lowest offers from Danubius (TL128) and Raaber (TL103) untrustworthy. Giesl lobbied on behalf of the Austro-Hungarian firms. He affirmed to the Turkish commission—at the request of the representatives of these firms and the Wiener Bankverein, which was undertaking the financial guarantee, and on the basis of the certification of the Austrian War Ministry—that both firms were efficient and had already delivered large portions of munitions wagons to the Austrian War Administration. The head of the Turkish commission, Major-General Hassan Riza Pasha, therefore decided to award the contract to Danubius and Raaber. Forced onto the defensive, Krupp had to cut prices to maintain the contracts for quick-fire guns. Ehrhardt, Skoda, Armstrong, and Schneider all bid higher prices than Krupp. For munitions, Krupp underbid Armstrong by 50 percent.[120]

On the eve of the First Balkan War, Krupp's monopoly in artillery finally cracked. Schneider secured a Turkish order for eighteen batteries (seventy-two guns) of quick-fire mountain guns in 1911, and the French firm followed with another contract in May 1912 for thirty-six guns. Then, as a result of trials for a light field howitzer carried out during 1912, the Turks signed a contract on May 6, 1913, with Skoda for thirty-six quick-firing howitzers (10.5cm) for 5.9 million kronen (£220,000).

Krupp still had clout in the Turkish market, though, as evidenced by its contract in June 1913 worth £700,000 for thirty-six quick-fire howitzers (15cm) and eight large howitzers (30.5cm) intended for Dardanelles.[121] In 1914 the Turks again ordered from Schneider a number of mountain guns as soon as the superiority of the French mountain gun to that of Krupp had been ascertained by their artillery expert, Riza Pasha.[122]

Joining in the process of artillery modernization, the South American countries now experienced their own version of the Krupp-Schneider artillery duel that was playing itself out in the Balkans. Although it involved comparable participants, the results turned out quite differently. As had been the case in the Balkans, the French tried to apply the financial assets of French banking to secure the artillery orders, whereas the Germans looked to manipulate through personal connections to the military ministers and the army officers.

Having been bested by Schneider in comparative gun trials in the Balkans in recent years, Krupp worked tenaciously to maintain its hold in South America. Here, too, Krupp was on the defensive. Argentine president Dr. Figuera Alcorta's address in May 1907 alluded to the trials of field guns being conducted on a range outside Buenos Aires. Krupp, Creusot, Ehrhardt, Armstrong, and Vickers-Maxim participated. Krupp guns sold to Argentina and Chile had proved very unsatisfactory, and had been dismissed in the Argentine press a few years earlier. In 1903 *La Prensa* had waged a campaign against Krupp artillery, and a deputy in Congress had brought an interpellation of the war minister. The Congress had voted 60–7 in support of the war minister, but the negative publicity had tarnished the firm's prestige since the gun tubes had proved very defective. In May 1906 a gunnery expert arrived in Buenos Aires in anticipation of Krupp's obtaining an order for new quick-fire field guns without the ordeal of trials.[123] No doubt the looming comparative trials gave Essen fits.

The prospect of two German firms in competition meant that the role of the German diplomats could prove decisive. Ehrhardt knew that the German Foreign Office favored Krupp. To neutralize the embassy, Rhein-Metall sent a letter to Wilhelmstrasse calling for cooperation among German firms. In the event of potential competition between German firms, Ehrhardt expressed its sympathy for working cooperatively. The firm requested that the Foreign Office, and especially the German ambassadors, take the initiative to coordinate a common German front in foreign orders. Krupp, however, had no desire to share with Ehrhardt. Krupp, in a letter to Bernhard von Bülow in November 1907, depicted the Argentine field gun trials as a classic contest between Germany and France. He reported that he had learned that the Argentine commission was divided into two camps: the larger group favored Krupp while the smaller group supported Schneider. According to Krupp, the minority hoped through the purchase of French guns to mitigate the further strengthening of German influence. Thus, concluded Krupp, Ehrhardt had already been eliminated, and the German ambassador should actively support Essen to ensure for German war industry the most work.[124]

When the gun trials commenced, the competition occurred on the field and off it. Krupp went first. The gun burst, and two other sample guns had to be used. The incident revived the complaints about the material supplied by Krupp. Undeterred by the mishap on the testing range, Krupp's agent had taken a house at Hurlingham immediately adjacent to the range. He entertained lavishly, spoke excellent Spanish, and was always ready to answer any questions. Armstrong's agent, in contrast, resided at the English Club at Hurlingham in a modest way, and without entertaining. His Spanish lacked the proficiency to enable him to give technical information about his gun. Vickers-Maxim's agent, Dardier, also kept a low profile, but he had many influential friends among the Argentines.[125]

Upon completion of the trials, the technical report on field guns was submitted to the minister of war, but contrary to custom, it was not published. Walter Townley's assessment of the situation contained much truth:

> The Argentine Government is in this matter between the devil and the deep sea; the German gun has the support of the Minister of War and of the clique of German-trained officers who surround him. On the other hand the French Minister threatens that, if the French gun is not given the reward it has fairly won, the stream of French capital to this country which is at present moment so warmly welcomed will dry up completely. In spite of assurances to the contrary, money from abroad will probably be wanted, if the armament scheme is approved by Congress, and there is reason to believe that hopes are entertained that it may be obtainable in Paris.[126]

For its army supplies, Argentina stuck with the Germans even though Krupp's gun had proved unequal to Schneider's. Despite the lack of officially published findings, word leaked that the report of the commission appointed to adjudicate had unanimously decided in favor of Schneider.

The Ministry of War, however, issued a decree bestowing the contract for eighty-seven batteries of field guns worth £2 million to Krupp. As the basis for its decision, the government pointed out that the army currently employed the Krupp system, and therefore the artillery personnel already had acquired familiarity with it. Obviously, the German instructors attached to the Argentine army, and the Argentine officers who had been trained in Germany, had preferred Krupp. No doubt, also, "the liberal presents distributed by Messrs. Krupp in which the highest officials concerned are believed to have participated"[127] did not hurt the German cause. During 1911 Argentina completed the rearmament of its artillery with Krupp M1909 quick-fire field guns. For the infantry the government bought 120,000 Mausers M1909 to replace older models.[128]

The French were livid. They deeply resented how the Argentine government had played them. The official reason given for selecting Krupp should have obviated the need for long and laborious trials. Schneider believed that the trials had only served to afford Krupp the chance to observe Creusot's system with an eye to remedying Krupp's deficiencies. The French government informed the Argentine minister in Paris that, under these circumstances, they would not permit money raised in France to buy German guns. As a reprisal, France prohibited the quotation of Argentine government securities on the Paris Bourse.[129]

For years German firms had been the sole providers of munitions of war to the Chilean army, but the Austrians worked more aggressively to reenter the market. Aware that Chile would likely purchase new weapons over the next few years, Baron Carl von Giskra, the Austro-Hungarian minister in Santiago in 1906, actively tried to advance Austrian commercial interests in Chile. Giskra brought a list of Austrian firms to the attention of President Pedro Montt and his Cabinet. Austrian hopes rose after an article in the press suggested that Chile should procure its new weapons from non-German sources, and pointed to Austria. The Germans reacted promptly to the threat. The German government sent a secret circular to German armaments firms warning about the Austrian competition and urging them not to lose the South American market. During his brief tenure in Santiago, Giskra failed to break the German dominance. He blamed this on the poor cooperation among Austro-Hungarian enterprises and Körner's still formidable pro-German influence in the Chilean army.[130]

The Austrians achieved some measure of success in the rifle business. Chile accepted tenders on rifle contracts in 1909–1910 for modernization of its Mausers. Steyr won the contract for 20,000 rifle stocks, 26,000 barrels, and 30,000 sights. The Austrian firm built on this modest gain and in 1910 won another contract, this time for new rifles and carbines. Under pressure from DWMF's supporters in the press, President Barros Luco withdrew the Austrian contract in 1911. Deutsche Bank and Disconto Gesellschaft provided Chile with a loan of 100 million marks to pay for a German order. However, in November 1911 DWMF informed the German Foreign Office that it could not handle the Chilean order for 30,000 Mausers because Argentina and Brazil had already placed larger orders for 100,000 rifles each, and the German plant had no capacity to spare for Chile. The German company offered to cut the Austrians in for a fee. In 1912 Chile concluded the contract with Steyr for 37,500 rifles and 5,600 carbines 1895M and 30 million cartridges for £400,000.[131]

In 1907 the Chilean government announced its plans to rearm the artillery by modernizing with quick-fire guns. The newspaper *Mercurio* on June 26 used the occasion to assert that since Krupp no longer held the monopoly on supplying foreign armies with weapons, the Chilean War Department should solicit tenders from various manufacturers in Europe. The press also advocated carrying out the necessary trials in Chile instead of sending a commission abroad, as had previously been the practice. The government anticipated acquiring artillery in the coming years to the tune of £1.2 million–£1.4 million (25 million marks). By 1909 newspapers were criticizing Körner's purchase of new artillery in Germany when Schneider's guns had proved superior in the Argentine gun trials. The Chilean gun trials, held in April, had excluded all firms but Krupp. Even though they faced no competition, the Krupp guns gave a poor showing. The official Chilean report voiced concern that in case of war, Chile could not count on Krupp artillery. The Chilean press also called attention to the defective state of the Chilean artillery. A great proportion of guns were unserviceable and the tubes expanded after 20–30 rounds of firing. Körner had ordered 80,000 rifles and 20 batteries of artillery in Germany.[132]

The question of the gun purchase, though decided in favor of Krupp, generated much controversy. The Chilean Artillery Commission in Germany had explicitly requested the import of Ehrhardt's guns for testing purposes. Nevertheless, the government had discarded the guns without allowing the trial. For years Ehrhardt had sought contracts in Chile, but to no avail. In 1912, Renato Valdes, a journalist known for his probing articles in *Diario Illustrado* against various abuses in government administration, agreed to work as Ehrhardt's agent. Valdes launched his anti-Krupp campaign by publishing a statement in the paper announcing his intention to donate his commission of 3 percent to charity in the event that Ehrhardt gained the contracts. His purpose was to expose Krupp's business methods, and to demonstrate that the Chilean military authorities were corruptly tied to Krupp to such an extent that they were willing to purchase inferior artillery at exorbitant prices.

Ehrhardt sent two of its guns for trials at the firm's expense. Krupp's agent and influential friends stopped the guns at the Los Andes custom house, saying that according to law no arms could be introduced into the country without the permission of the government. Public opinion eventually prevailed after Senator Walker Martines raised questions in the Senate. The government relented and decreed that Ehrhardt's guns could be sent on from Los Andes. Krupp's agent then arranged to allow the gun but not the ammunition or spare parts. This was overruled by government decree. It became known in the Senate that the Artillery Commission in Berlin preferred Ehrhardt guns, but that the Ministry of War insisted on Krupp. Given Krupp's tremendous influence, no one expected an impartial verdict on the Ehrhardt guns. RheinMetall had achieved a moral victory by getting to the trials at all.[133] As Lowther observed, "The complete Germanization of the Chilean army may preclude the idea of purchasing guns in France, or elsewhere, but the rival merits of different German firms at least may now be taken into consideration."[134]

In Brazil Krupp maintained its position through mysterious circumstances and devious actions. In 1902 Schneider had sent to Brazil one of its new model quick-fire 75mm artillery pieces for comparative trials against the Krupp gun. In objective technical terms, the Schneider gun outclassed the Krupp piece. The trials had not concluded before the Brazilian Artillery Commission requested that the tests recommence with new material. Consequently, in 1903 Schneider provided another piece. The new test gun perished under strange circumstances when a mob set fire to its storage facility and the gun barrel was made useless. Krupp accused the French of having destroyed their own gun to avoid competition, but the French blamed the Germans for sabotage. The gun trials were now delayed until 1904. The French again sent their material, but the shipping agent, in all likelihood bribed by Krupp, refused to unload the shipment because it contained explosive materials. After interminable meetings and delays, a nasty press campaign followed in which the Germans asserted that Schneider had armed Peru in that country's hostile designs against Brazil. In this charged atmosphere, Brazil gave the order to Krupp. However, Schneider did sell Brazil some coastal guns in 1911.[135]

In Eastern Europe during the critical decade prior to World War I domestic and foreign pressures combined to politicize the arms trade to an unparalleled degree. Domestically, the heightened politicization of the procurement process led to cabinets falling and reshuffling. The award of contracts and loans, along with published results from the firing ranges, became political fodder in the debates and interpellations on the floors of the national parliaments and were splashed across the headlines of domestic newspapers. The intensifying competition on the part of the foreign firms amplified the pressure. The Germans and Austrians found themselves on the defensive as they tenaciously fought to stave off the charge from the French. The industrial competition drew in the diplomatic resources of the supplier governments, usually with detrimental results for the companies involved. The Austrians tried to coerce the Serbs, and the Germans insulted the Greeks. Ultimately, in the artillery duels the French overpowered their rivals with stronger financial means and a technically superior product. These armaments acquisitions did not constitute an arms race. Rather, the purchases formed part of a general force modernization that had essentially concluded by about 1908–1909. While it is true that Serbia engaged in a frantic search to purchase as many as 400,000 new rifles in 1914, the Serbian moves should not be seen as an arms race. They reflected the need to replace worn-out stocks and to expand the Serbian army commensurate with the doubling of the country's territory and population as a result of victories in the two Balkan wars. Having been frustrated by the French War Ministry's reluctance to tie up the state plant at Châtellerault for a five-year production run, the Serbs had eventually signed a contract with Vickers for 208,000 rifles in July 1914, but the outbreak of World War I made the arrangement a dead issue.[136]

The question remains: why did such different outcomes arise for

Krupp and Schneider in South America given that the same guns competed there as in the Balkans? In South America the French threatened governments with the stick of financial embargo if the orders did not go their way, while the Germans offered the carrot of insider profits to keep the French out. In spite of the technical superiority of its artillery, Schneider lost out to Krupp in South America due to stronger forces of corruption. In both regions the procurement processes involved firing tests, professional deliberations, and political calculations. Whereas in the Balkans the technical tests generally proved decisive in establishing the merits of the Schneider guns, in South America Krupp's relationship with the pro-German officers succeeded in preempting field tests entirely or having poor tests overlooked. In neither region did the foreigners succeed in forcing arms on the buyers through corruption. The desire to acquire the arms in the first place came from within the customer countries. Corruption led to South American states buying inferior guns in the case of Krupp over Schneider.

Conclusion

Distinct national differences in attitude toward the arms business emerged on the part of Great Power governments. Britain offered the least support for arms sales diplomatically and financially. The British government through its diplomats often voiced opposition to naval and military sales in the Balkans and South America even when British firms benefited. The British government's lack of enthusiasm for the arms trade derived partially from a consistent laissez-faire attitude that left the private firms to their own devices and that deemed all these expenditures by foreign governments a colossal waste of resources. With remarkable regularity British ambassadors, military attachés, and even naval officers employed by foreign states all tried to restrain the appetites of their host countries for more and bigger armaments purchases. London's desire to preserve Britain's naval dominance in the world probably colored their views. Any major expansion of the world naval trade brought with it a corresponding diminution of the Royal Navy's quantitative advantages, even if British shipbuilders claimed the lion's share of the business. In South America the British government had an interest in defusing the naval race between Chile and Argentina for economic reasons, and London rarely encouraged the sales by British firms. In East Africa the British vainly sought to curtail the rifle trade, but they watched helplessly as Italian and then French suppliers flooded Ethiopia with arms. Only Japan proved a notable exception to the general British pattern. The British government actively intervened to help Japan and hinder Russia in 1903. However, the British government did not expend its efforts to win Japanese naval contracts for British firms. Instead, British motivations stemmed from a strategic interest to contain Russia in Asia, and to support Japan, a formal ally by treaty since 1902.

German, French, and Austrian officials, including diplomats and armed forces officers, played more aggressively compared to their British counterparts. For the continentals, lobbying for armaments orders comprised an important part of their duties. The more active role taken by German, French, and Austrian officials only partially accounted for their success over the British, or Americans, for sales of army equipment. National reputations built by successes on the battlefield played as great a role, if not greater. American rifles sold themselves after the Civil War without the activist intervention of the U.S. government. Similarly, the British, as the recognized leading naval power in the world, consistently dominated the naval trade regardless of British diplomats' admonitions against the purchases by foreign states. The French found themselves in second place for artillery and naval sales, and the Austrians earned second place for rifle sales. In these instances French and Austrian firms had won their positions largely based on technical qualities and financial support rather than effective diplomatic lobbying.

The Germans held the top spot for military prowess based on their victory over the Austrians and the French by 1871. The reverberations from these Prussian victories could be felt around the globe as the Germans became the standard model for modernizing armies. Of still greater significance for Krupp's sales worldwide, the Russo-Turkish War, 1877–1878, proved the superior potency of German artillery even in the hands of non-Germans. German-made weapons now gained markets as importer states considered them an important element of the German system emulated by other countries. The Prussianization of the Chilean army during the 1890s was but one example. In southeastern Europe, South America, and East Asia buyer countries embraced German hardware for their armies. Krupp stood head and shoulders above all others as the number one artillery supplier, having established itself as the predominant supplier for these regions by 1880. The German military missions in the Ottoman Empire (1882) and Chile (1891) only reinforced the preexisting trend. However, the establishment of a domestic artillery industry in Russia removed one of Krupp's biggest customers. To compensate for the loss of the huge Russian market, Krupp aggressively looked to South America and East Asia with the result that Krupp's sales to those two regions dramatically surpassed Eastern Europe in the years 1880–1904.

Based on the detailed studies of the military in Chile and Turkey, one could easily conclude that German army officers working in concert with diplomats and the German armaments firms dictated arms sales to the buyer states and wielded sufficient influence to Prussianize these countries. In a global context, though, the role of German military officers as principal agents of the arms business should not be overstated. Certainly German officers did advocate in favor of German sales. Significantly, the two areas where German officers were directly involved in armaments acquisitions, Chile and the Ottoman Empire, were not the biggest customers in the period 1880–1904. Japan bought the most Krupps worldwide, and Argentina bought twice as many as Chile. Even the Turks bought less during the years of the German Military Mission than in the preceding era. Furthermore, upon integrating the naval trade into the picture, the perception of overwhelming German dominance and influence diminishes dramatically. The Chilean navy bought almost exclusively from British firms. Indeed, the global market remained primarily British-dominated, although in Eastern Europe the Turks and Russians bought Schichau ships.

Like other Great Powers, the tsarist government in Russia considered arms sales as a means to advance political purposes. However, Russia as an arms exporter differed from the other European suppliers in important ways. First, whereas the tsarist government directly engaged in sales negotiations and concluded contracts, in other European states this role fell to private firms. Related to the absence of any Russian private arms exporter, the Russian foreign arms sales were decidedly not driven by profit. Finally, the tsarist government turned out to be an ambivalent exporter, and in part the potential success of Russian sales to the Balkans was undermined by political cautiousness and conflicting priorities between the Foreign Ministry and the War Ministry. Russia could not easily spare arms for sale abroad until its own military-industrial needs were met.

As a generalization, the supplier countries poorly integrated the arms business into diplomatic strategy. Transregional trends overrode diplomatic arrangements as British ships and German artillery prevailed in the global arms business in the period 1880–1904. Generally, the periodization for the suppliers did not track closely with the diplomatic periodization. Krupp's and Mauser's hegemony preceded German *Weltpolitik* rather than being caused by it. The triumph of Krupp sales in

East Asia and South America resembled those in the Balkans, even though the non-European states held no place in the Bismarckian diplomatic system. The Franco-Russian military alliance by 1894 certainly had an effect on the Russian arms trade, since it is hard to imagine the Russians placing rifle orders in a French state factory without it. Yet the intense Anglo-Russian rivalry prior to 1907 did not lead to the exclusion of the English from the Russian naval market as contractors for machinery and technical services. Moreover, U.S. firms did as well as the French in naval sales without having any alliance with Russia. The Franco-Russian alliance possibly helped the French in the Balkans and to a lesser extent in Ethiopia, but it did not preclude Russian dealings with Germans and Austrians. As one moved farther away from Western Europe, the diplomatic connections became even less relevant. Often Krupp acted more boldly than the German government, as in China and the Ottoman Empire. The Anglo-Japanese alliance brought the Japanese some British help aimed against Russia, but British firms would have done well anyway because of the reputation of the Royal Navy as the best in the world.

From 1904 to 1914 the key changes in the international system centered on the evolving entente. In 1904 Britain and France came together into the entente cordiale, followed a few years later by the Anglo-Russian Convention of 1907. These new diplomatic arrangements, combined with the now well-established Franco-Russian Alliance, potentially brought together Britain, France, and Russia as a rival bloc to the Austro-German Dual Alliance. One could easily interpret these various business alliances and the Schneider-Krupp competition as simply an economic manifestation of the Franco-German diplomatic rivalry and the larger antagonism between the Triple Entente and the Dual Alliance. Besides the problem of reading history backwards from 1914, such a reductionist approach is unsatisfactory, as it suffers from a serious misreading of the autonomy of the business interests participating in the arms trade. The armaments firms used the diplomats for their private gain far more successfully than the diplomats used the firms for the furtherance of national objectives.

In any case, bankers showed themselves more persuasive than ambassadors when it came to negotiating a contract and closing a deal, and the private companies exhibited a greater willingness to cooperate in groupings that transcended the boundaries of the alliances. The Germans actively participated in the Russian business despite growing anti-German sentiments in Russian foreign policy. In 1912, with the approval of the

German government, Schichau actually established production facilities within the tsarist empire near Riga. In appreciation of this investment, Russia ordered two small cruisers of 4,350 tons' displacement from the Schichau branch in Danzig. By the end of 1913 the Riga works entered production, and the tsarist government granted the plant contracts for nine more torpedo boats.[1] On the eve of World War I the Russian Putilov Company had arranged direct investment projects with Skoda, Schneider, and the German firm Blohm und Voss.

Additionally, there was as much rivalry and competition between firms from within the same alliance as from without. The arms trade in Ethiopia serves as a good example of the problem. Regardless of the Anglo-French alliance, the French arms trade from Djibouti to Ethiopia flowed on in spite of repeated British complaints and requests for French cooperation in halting it. Likewise, naval rivalries between British and French firms for sales to Greece intensified. Similarly, the Austro-German Dual Alliance since 1879 did nothing to alleviate an intense rivalry in the Balkans and South America between Krupp and Skoda or DWMF and Steyr, and Vienna's laments fell on deaf ears in Berlin.

Firms may have presented their export sales as helping secure national interests for their home governments by securing "influence" in the buyer countries, but no strong link existed between the arms trade and alliances with the buyer countries. The arms trade did not give supplier states controlling influence based on the armaments alone. Clearly the application of influence in the negative through pressure and threats rarely worked. Cases showed remarkably little leverage when Austria tried to strong-arm the Serbs, the Russians tried to cut off the Bulgarians, or the French tried to subject the South American states to financial blackmail. In fact, these attempts backfired. However, the exercise of positive influence had only limited effect. The generous supply of arms from Russia to Romania before 1878 did not prevent Romania from an anti-Russian orientation in the following era.

In East Africa, the arms trade was intimately bound up with imperialism. Maintaining and controlling an arms trade supply line had provided a key element in Menilek's building of the Ethiopian state. By clever diplomatic maneuvering he had beaten back Italian imperialism with the very weapons they had delivered into his hands to influence him. At the same time, the allure of the Ethiopian arms trade prompted the French to establish a presence in Djibouti, and from the very beginning that colony's fate was linked to the rifle trade with Harar. Menilek's subsequent embrace of the French arms trade preserved his military power. It also caused the British much consternation, since a central goal for British imperialism in the region was to curtail the arms business out of fear that it would enable resistance and threaten British possessions.

The practices of the armaments firms could be classified as imperialistic, albeit of an informal variety. Within the independent buyer states the companies strove to control an aspect of domestic policy, namely procurement. They sought out and cultivated military and naval officers, politicians, and even monarchs as collaborative partners in the acquisition process to lock up these countries as informal "arms trade colonies" and block out alternative suppliers. Krupp worked most consistently and effectively in this manner, but other firms tried as well.

The aggressive tactics of the industrial firms and their banking junior partners did more in facilitating arms sales than the diplomats. Firms engaged in forms of business diplomacy with one another and with the banks. Price fixing and market sharing lay at the heart of the international rifle cartel, and these arrangements offered a way to manage the Steyr-Mauser competition. The Vickers-Armstrong amalgamation forged a formidable team, and they worked to secure a monopoly supplier position through their naval docks contracts in the Ottoman Empire. The British pair also entered into an alliance with Schneider to combat Krupp in Turkey. In East Africa the French colony of Djibouti actually functioned as a formal "arms trade colony" since the revenue from the arms trade funded the colonial administration, and French colonial administrators personally participated in the rifle trade.

Why did buyer states behave as they did? In the emerging world system of the armaments trade the domestic political policies and choices of individual polities could result in import dependence or domestic self-sufficiency. These choices had ramifications for taxation, debt, and development or underdevelopment. The arms makers did not cause the underdevelopment of buyer countries, nor did they force the buyers to rely on importing the hardware rather than domesticating the technology. Indeed, the foreign firms willingly transferred the technology instead of forcing states to import the finished defense goods and preempting the development of domestic war industries. Rather than any conscious design by foreigners, the process leading to the rise of the international arms trade should be attributed more accurately to the interplay of external and internal factors. Externally, the impersonal forces of rapid technological change and the development of an international armaments mass market made the rapid acquisition of the latest defense equipment a financially sound choice. Internally, the political and personal decisions of the policy-makers profoundly affected the course of the armaments policy. The greater emphasis should be placed on domestic agency over foreign manipulation as the key to understanding the growth of arms imports.

But did technical and professional considerations win out over corruption and bribery? In the Ottoman system, especially under Abdulhamid II, corruption and attempts to use the placing of orders as a means to purchase diplomatic backing acted as driving forces. In Argentina, Brazil, and Chile corruption overrode professional and technical considerations for military orders after 1905, while non-British naval orders were subjected to more political/diplomatic considerations. In Chile the personal preferences of President Balmaceda for France overrode professional naval preference for British ships. Chilean officials grew frustrated with the French, and in the post–civil war era turned back to England. The personal preference of Admiral Montt kept Chile a preserve of British naval orders. In the Balkans, financial/technical reasons dominated, except in Romania, where personal-political involvement by the king determined the outcome. In China bribery played some part; in Russia technical issues dominated with some incidental bribery. Ethiopia did not develop a technical or professional procurement process. Japan proved the most meritorious in its procurement professionalism.

The arms trade generated foreign debt through loans, and it connected directly to domestic taxation policies. Japan, Russia, and China largely paid for their armaments through their own monies. Ethiopia used its trade resources of gold, ivory, and civet. Argentina, Brazil, Chile, the Ottoman Empire, and the Balkan states turned to foreign loans. Without willing and able financial groups, contracts would not have gone through. Turning to loans inhibited domestic industrial development in buyer states because money effectively stayed in Europe, transferred from European banks to European firms. Furthermore, although many states failed to achieve self-sufficiency in armaments production and others did not even try, the buyer states did obtain high-quality equipment quickly and relatively cheaply given their limited financial means. Through the creation of excess capacity and the availability of Western financial bond and banking markets, private enterprise created a buyers' market and a borrowers' market. Yet as the business competition intensified the armaments firms themselves increasingly offered to provide modernized plant and technical aid within the buyer countries.

Explanations for armaments and force levels not based on rational calculations but derived from culture, religion, or ideologies do not hold up well for the years 1860–1914. For example, James Payne has argued for the post-1945 period that orientations toward military force primarily rooted in Islam or Marxism essentially explained the higher-than-average military efforts for Ethiopia, Somalia, Turkey, Bulgaria, the Soviet Union (Russia), and Yugoslavia (Serbia). Along with those states Greece and Chile also ranked above average and in the top forty.[2] Yet we have seen that the Eastern European states were already leading the charge for armaments in the precommunist period, and therefore it seems unlikely that the advent of socialist systems in and of itself caused the preference for more armaments. Likewise, the idea that Islam caused the Ottomans to purchase large quantities of arms does not stand up to scrutiny, and East Africa from the Somali coast to Ethiopia was already flooded with arms by the early twentieth century. These cases reveal the merits of the qualitative approach over a purely quantitative one and highlight the critical importance of peering into the black box of motivations, perceptions, and misperceptions of the participants at the time.

A review of the buyer states examined here lends qualified support to the domestic structure model by revealing some connection between the degree of democratization in a given country and its degree of acquisition of armaments. The states covered offer a range of polities, from autocracies to constitutional monarchies to nascent republican democracies. The less autocratic did not have a lesser propensity toward armed competition. On the contrary, when mass politics entered the procurement process, a degree of democratization invigorated the process. The transition from autocracy to some form of parliamentary system paved the way for armaments acquisition to be used as an expression of the national will and national vitality. It is at this level that the transition from arms trade to arms race happened. Thus, the advent of constitutional systems and the requirements for legislative approval for arms funding did nothing to discourage massive armaments purchases in Japan, the small Balkan states, or Russia and the Ottoman Empire after their respective revolutions. The South American republics similarly plunged into two naval races with congressional support. For the buyer states, national self-defense readily changed into military and territorial aggrandizement. Japan offered the most striking example of arms importer turned imperial power. The Balkan states too quickly changed from arming in the name of national independence to building up armed strength to expand their borders at the expense of the Ottoman Empire and each other. In South America, Chile led the charge for regional dominance, to be followed by Argentina and then Brazil.

The motivations, and correspondingly the meaning of the arms trade for the buyer states, ranged from defensive to offensive. Without a doubt the same armaments potentially could be used aggressively or defensively, and therefore characterizing a set of purchases as defensive or offensive can prove problematic. Nevertheless, one can distinguish between offensive and defensive weapons acquisitions primarily based on buyer motive as determined by internal policy discussions. A buyer state's desire for territorial aggrandizement can reflect an offensive armaments posture, whereas the desire to preserve the status quo can be considered defensive armament. However, even defensive armament could be misinterpreted by neighboring states as offensive in nature and trigger an action-reaction arms race.

Defensive motives for the restoration of previous status and efforts at self-preservation drove the Ottomans, Chinese, Russians, and Brazilians. Among these powers, the Turks and the Chinese found themselves in roughly analogous positions, since both lacked time and money throughout the period and each had to face wars and only intermittent peace from regional and great powers. There was not enough time to wait around for domestic supply capabilities to come on line. To preserve its Great Power status, Russia emulated the military-production capabilities of the major international war suppliers. Russian domestic producers continued to depend on domestic contracts from their own government rather than insinuating themselves into the export market. Despite its impressive and rapid industrial gains, Russia occupied a far more important place in the international armaments trade as an importer rather than an exporter. The Russian government imported not only the defense hardware, but the technology as well. The Brazilians embarked on their naval race after 1905 with the intention of restoring the pride of place they had occupied in the region during the past.

Nationalism and imperialism served as offensive motivations. For the

Balkan states the road to national recognition ran through armed conflict, and therefore the arms trade was tightly bound with assertions of national sovereignty and independence. Kingly rivalries and state-building motivated the arms trade in Ethiopia. In South America the motivation to climb up the ladder of regional prestige and predominance pushed Chile and Argentina, but their naval race managed to avert war. On the offensive side, Japan moved from self-strengthening to Great Power aspirations. The Japanese let armaments drive fiscal policy. Unlike the Ottomans or the Chinese, the Japanese had the luxury of choosing the time and place of their offensive wars, and this enabled them to make long-range plans for armed industrial modernization. On this count, the Japanese and Russian approaches resembled each other as Japan, too, domesticated the technology. While in world history it is common to focus on Japan's methodical development of domestic industry as the key to its rising power, native industry alone could not have made it possible. The arms trade figured more prominently in the short run, and it proved an essential element for Japan to enter the exalted ranks of the imperialist powers.

The Russo-Japanese War marked a watershed for Japan in the international arms trade. Increasingly, Japan not only supplied itself, but also began to figure as an arms exporter. Following the Japanese victory in 1905 the rising Asian power successfully entered the foray into naval exports to China and exports of captured Russian surplus rifles to Ethiopia. In the period 1884–1903 almost 90 percent of Japan's navy had been imported, whereas between 1904 and 1921 imports dropped to 13 percent. Instead of importing, Japan formed a joint venture with Armstrong and Vickers. The new Japanese company, capitalized with £1 million, half supplied by Japanese government and half by Armstrong and Vickers, manufactured war material in Japan. The Japanese government placed orders with this company for all war materials they could not make for themselves in their own dockyards and arsenals. With the help of Vickers, steel production in Japan soon doubled and private capital moved into new industry to supplement government efforts. Japan still imported most building materials by 1914. As arms exporters, the Japanese unsuccessfully entered into the naval sales competition in South America and managed to sell Russian surplus rifles to Ethiopia after 1910. They had better fortunes in China. The Chinese ordered six gunboats from the Kawasaki Dockyard in Kobe in 1904. From 1906 to 1907 China purchased seven more gunboats from Kawasaki. Although the Japanese remained unsuccessful in their bid to gain sales to South America, the South American states took notice of the changed naval environment.[3]

During the first decade of the twentieth century the global arms trade transformed from a series of discrete regional components into a globally interactive system. The conclusion of the first South American naval race (Chile and Argentina) in 1902 helped fuel the Japanese-Russian competition in such a way as to form a Pacific Rim naval circuit. The second South American naval race (Brazil-Argentina-Chile) generated surplus dreadnoughts for the Greco-Turkish naval race and indirectly made an impact on Russia. Meanwhile, the results of the intensive artillery tests conducted by the Balkan states became well known to the South American states. Furthermore, the success of the Balkan states armed with French artillery over the Turks and their Krupp guns during the First Balkan War forced the Germans onto the defensive in their South American markets.

Taken together, military and naval imports played an enormous role in equipping the armed forces in Eastern Europe, East Asia, and South America. In turn, the armaments firms themselves came to rely on the exports to these regions as part of their overall sales. During peacetime, the major private defense producers could not count on sufficient business from their home governments, and as a result exports proved vital to the firms' prosperity. The domestic structure model and the action-reaction model have proved more accurate on the buyer side, but their state-centric focus omits the business dimension from the analysis. Although the two models in complementary fashion do much to explain why state leaders might pursue armaments, they do not offer any insight into why a given government chose to buy from a particular supplier. On this point, the business side comes into play. Krupp, Mauser, Steyr, Schichau, and Schneider all found huge stakes in Eastern Europe. Armstrong, Yarrow, and Laird along with other British firms dominated in China, Japan, Argentina, Brazil, and Chile. Given how Eastern European states consistently served as big customers, it would be no exaggeration to conclude that for the global arms trade Eastern Europe ranked as the driving region from 1860 to 1914.

From within Eastern Europe the Russian factor acted as a significant catalyst for the arms trade across Eurasia. David Stevenson and David Herrmann have convincingly shown the role of Russian rearmament after 1909 as a principal cause of the land arms race in Europe leading to World War I.[4] Consideration of the problem of Russian rearmament after 1909 in light of the previous half century illuminates the pivotal role Russia played, directly and indirectly, in increasing its neighbors' armaments. Thus, Russia's defensive rearmament and modernization, according to its own lights, were perceived as offensive by its neighbors. The Russo-Turkish rivalry served as the central defining relationship for the arms trade in Eastern Europe. By the time of the Crimean War, the technological developments of industrialization threatened to reduce Russia's stature to the level of the Turks.

For the Russians after 1856, armaments meant the preservation of their prestige as a Great Power in the European states system. While the central tension between Russia and the Ottomans in the eastern question had functioned as the engine of the arms trade in Eastern Europe, Russian expansion also stoked the global arms trade at the other end of Eurasia. Russian imperialist expansion into Central Asia accelerated Chinese imports after 1874 and stimulated the Japanese to become the leading buyer state in East Asia in the second half of the 1890s. The Japanese government had determined on imperial expansion back in the 1880s, but the scale of the purchases reached such staggering proportions after 1895 because the race to match Russia, the largest power in the world, demanded extraordinary commitment of resources. Following its defeat at the hands of Japan in 1905, Russia again confronted the challenge of rearmament as a prerequisite for preserving its Great Power status. As a European Great Power, Russia's position proved critical for the stability and peace of the European states system. Within Russia the arms trade intersected with rearmament and imperialism. Just as Russian territorial expansion and the size of Russian armed forces had facilitated arms races at the eastern end of Eurasia for China and Japan, Russian rearmament profoundly affected western Eurasia after 1909 and significantly contributed to the road to war in Europe in 1914.

Notes

Abbreviations

ADM	Admiralty Office, British National Archives (formerly Public Record Office)
AJP	Archibald Johnston Papers
BDFA	British Documents on Foreign Affairs. Kenneth Bourne and D. Cameron Watt (eds.), *British Documents on Foreign Affairs— Reports and Papers from the Foreign Office Confidential Print, Part I, From the Mid-Nineteenth Century to the First World War* (Frederick, MD: University Publications of America, 1984).
FO	Foreign Office, British National Archives
GFM	German Foreign Ministry microfilm, British National Archive
GP	*Die Grosse Politik der Europäischen Kabinette*
HHSA	Haus-, Hof-. Und Staatsarchiv, Vienna
KA	Kriegsarchiv, Vienna
KAE	Krupp Archive, Essen
MAE	Ministrère des affaires Etrangères, Paris
RGVIA	Russian State Military Historical Archive, Moscow
RIHS	Rhode Island Historical Society
SA	Schneider Archive, Le Creusot
SAP	Skoda Archive, Plzen
TWAS	Tyne and Wear Archive Service, Newcastle-on-Tyne
VHS	Virginia Historical Society
WO	War Office, British National Archives

Introduction

1. David Stevenson, *Armaments and the Coming of War: Europe, 1904–1914* (Oxford: Clarendon Press, 1996); David G. Herrmann, *The Arming of Europe and the Making of the First World War* (Princeton, NJ: Princeton University Press, 1996); Peter Gatrell, *Government, Industry and Rearmament in Russia, 1900–1914* (Cambridge: Cambridge University Press, 1994); Paul G. Halpern, *The Mediterranean Naval Situation, 1908–1914* (Cambridge, MA: Harvard University Press, 1971); Milan N. Vego, *Austro-Hungarian Naval Policy, 1904–1914* (London: Frank Cass, 1996). For Russia, see also J. N. Westwood, *Russian Naval Construction, 1905–1945* (London: Macmillan Press, 1994).

2. Barry Buzan and Eric Herring, *The Arms Dynamic in World Politics* (Boulder, CO: Lynne Rienner, 1998); Grant T. Hammond, *Plowshares into Swords: Arms Races in International Politics, 1840–1991* (Columbia: University of South Carolina Press, 1993); James L. Payne, *Why Nations Arm* (Oxford: Blackwell Press, 1989); A. F. Mullins Jr., *Born Arming: Development and Military Power in New States* (Stanford, CA: Stanford University Press, 1987).

3. Jonathan A. Grant, *Big Business in Russia* (Pittsburgh, PA: Pittsburgh University Press, 1999), 54–56; Robert J. Winklareth, *Naval Shipbuilders of the World from the Age of Sail to the Present Day* (London: Chatham, 2000), 21, 94–95; Theodore Ropp, *The Development of a Modern Navy: French Naval Policy, 1871–1904* (Annapolis, MD: Naval Institute Press, 1987), 63–65; William Manchester, *The Arms of Krupp, 1587–1968* (Boston: Bantam Books, 1968); Willi Boelcke, *Krupp und Die Hohenzollern, aus der Korrespondenz der Familie Krupp, 1850–1916* (Berlin: Rütten & Loening, 1956); Bernhard Menne, *Blood and Steel: The Rise of the House of Krupp* (New York: Lee Furman, 1938); Basil Collier, *Arms and the Men: The Arms Trade and Governments* (London: Hamish Hamilton, 1980); Zdenek Jindra, *Der Rüstungs-Konzern Fried. Krupp AG., 1914–1918* (Praha: Univerzita Karlova, 1986); Clive Trebilcock, *The Vickers Brothers: Armaments and Enterprise, 1854–1914* (London: Europa Publications, 1977); Edward R. Goldstein, "Vickers Limited and the Tsarist Regime," *Slavonic and East European Review* 58, no. 4 (1980): 561–571; Claude Beaud, "De L'Expansion Internationale a La Multinationale Schneider En Russie (1896–1914)," *Histoire Economie et Societe* 4 (1985): 575–602; Dietrich Geyer, *Russian Imperialism: The Interaction of Domestic and Foreign Policy, 1860–1914* (New Haven, CT: Yale University Press, 1977), 255–272.

4. The literature on imperialism is vast. For a good overview of metrocentric, systemic, and pericentric schools of thought, see Michael W. Doyle, *Empires* (Ithaca, NY: Cornell University Press, 1986), 19–47, 123–138.

5. Grant, *Big Business,* 71–72; Bruce I. Gudmusson, *On Artillery* (Westport, CT: Praeger, 1993), 21; Herrmann, *Arming of Europe,* 17; Stevenson, *Armaments and the Coming of War,* 17.

6. David C. Evans and Mark R. Peattie, *Kaigun: Strategy, Tactics and Technology in the Imperial Japanese Navy, 1887–1941* (Annapolis, MD: Naval Institute Press, 1997), 53–56.

7. Dick Keys and Ken Smith, *Down Elswick Shipways: Armstrong's Ships and People, 1884–1918* (Newcastle: Newcastle City Libraries, 1996), 7, 46–47.

8. *100 Jahre Schichau, 1837–1937* (Berlin: VDI-Verlag, 1937), 32–33, 38, 71–72; Eberhard Westpahl, *Ein Ostdeutscher Industriepionier: Ferdinand Schichau in sienem Leben und Schaffen* (Essen: West Verlag, 1957), 74–79.

9. Ropp, *Development of a Modern Navy,* 63–69.

10. Gary E. Weir, *Building the Kaiser's Navy: The Imperial Navy Office and German Industry in the Von Tirpitz Era, 1890–1919* (Annapolis, MD: Naval Institute Press, 1992), 12–16, 18, 29–30.
11. Keys and Smith, *Down Elswick Shipways*, 46–47.
12. Winklareth, *Naval Shipbuilders of the World*, 28.

1. Arsenals of Autocracy

1. Geoffrey Shannon Stewart, "The American Small Arms Industry: Its Search for Stability, 1865–1885" (master's thesis, Brown University, 1973), 64–65.
2. David Pam, *The Royal Small Arms Factory, Enfield and Its Workers* (Enfield: Jubille Hall, 1998), 73–74; David Stevenson, *Armaments and the Coming of War, Europe, 1904–1914* (Oxford: Oxford University Press, 2000), 16; Joseph Bradley, *Guns for the Tsar: American Technology and the Small Arms Industry in Nineteenth-Century Russia* (Dekalb: Northern Illinois University Press, 1990), 99, 102; R. W. Beachey, "The Arms Trade in East Africa in the Late Nineteenth Century," *Journal of African History* 3, no. 3 (1962): 452.
3. Roger Owen, *The Middle East in the World Economy, 1800–1914* (London: University Paperback, 1987), 59–62. Portions of following Ottoman discussion here and in subsequent chapters first appeared in Jonathan Grant, "The Sword of the Sultan: Ottoman Arms Imports, 1854–1914," *Journal of Military History* 66 (January 2002): 9–36.
4. FO 78/1508, Bulwer to Earl Russell, 4 July 1860.
5. Afif Büyüktugrul, *Osmanli Deniz Harp Tarihi III: Cilt* (Istanbul: T. C. Deniz Basimevi, 1973), 1; Justin McCarthy, *The Ottoman Turks* (New York: Addison Wesley Longman, 1997), 301–304.
6. Erkem Mustevellioglu, *Osmanli Askeri Te kilat ve Kiyafetleri, 1876–1908* (Istanbul: Askeri Müze ve Kültür Sitesi Komutanligi Yayinlari, 1986), 9–10.
7. FO 78/2216, Rumbold to Granville, 3 April 1872.
8. FO 78/2383, Elliot to Earl of Derby, 3 May 1875.
9. FO 78/2391, Elliot to Derby, 14 Dec. 1875.
10. ADM 231/10, Report 127, Turkish Fleet and Dockyards 1886, 3–4.
11. Nejat Gülen, *Dünden Bugüne Bahriyemiz* (Istanbul: Kastas A. S. Yayinlari, 1988), 122.
12. FO 78/1507, Bulwer to Russell, 13 June 1860; FO 78/1276, Alison to Earl of Clarendon, 27 Dec. 1857; Büyüktugrul, *Osmanli Deniz*, 16–18.
13. FO 78/1578, Bulwer to Russell, 13 Nov. 1861; ibid., Erskine to Russell, 14 Sept. 1863; FO 78/1783, Erskine to Russell, 9 Sept. 1863.
14. FO 78/1803, Bulwer to Russell, 19 May 1864.

15. FO 78/1804, Bulwer to Russell, 4 June 1864; ibid., Bulwer to Russell, 20 June 1864.

16. FO 78/1806, Bulwer to Russell, 30 Aug. 1864.

17. Ibid.; FO 78/2178, Elliot to Granville, 17 Nov. 1871; A. Gallenga, *Two Years of the Eastern Question*, vol. 1 (London: Samuel Tinsley, 1877), 248–249; Roger Chesneau and Eugene M. Kolesnik, eds., *Conway's All the World's Fighting Ships, 1860–1905* (New York: Naval Institute Press, 1976), 388.

18. FO 78/2178, Elliot to Granville, 17 Nov. 1871.

19. N. Grechaniuk, V. Dmitriev, F. Krinitsyn, Iu. Chernov, *Baltiiskii flot* (Moskva: Voennoe Izd. Ministerstva Oborony Soiuz SSR, 1960), 100.

20. FO 65/1141, Russia Commercial Report. Arthur Herbert, St. Petersburg, 22 Nov. 1882: enclosed in Thornton to Granville, 22 Nov. 1882.

21. Grechaniuk et al., *Baltiiskii flot*, 101; Edmund Ollier, *Cassell's Illustrated History of the Russo-Turkish War*, vol. 1 (London: Cassell Pelter and Galpin, 1880), 134.

22. Oral Sander and Kurthan Fişek, *Türk-ABD Silah Ticaretinin Ilk Yüzyılı, 1829–1929* (Istanbul: Erdini Basım ve Yayimevi, 1977), 56.

23. FO 78/2122, Elliot to Clarendon, 25 May 1870.

24. FO 78/2125, Elliot to Grenville, 6 Oct. 1870.

25. Felicia Johnson Deyrup, *Arms Makers of the Connecticut Valley: A Regional Study of the Economic Development of the Small Arms Industry, 1798–1870*, Smith College Studies in History (Northampton, MA: George Banta, 1948), 210–212; Herbert G. Houze, *Winchester Repeating Arms Company: Its History and Development from 1865 to 1981* (Iola, WI: Krause Publications, 1994), 75, 97, 107; RIHS, Box 3, folder 13, Joseph K. Ott, "Providence-Made Gun a Victim of Intrigue after Civil War," *Providence Sunday Journal*, 6 Feb. 1977.

26. RIHS, Providence Tool Company Records, Box 3, folder 1, telegram, Schuyler Hartley Graham to J. B. Anthony, 9 May 1872; Providence Tool Co. to Saint Saurent, February 1872.

27. RIHS, Providence Tool, Box 3, folder 1, J. A. Edwards to J. B. Anthony, 8 July 1872.

28. RIHS, Providence Tool, Box 3, folder 1, John B. Anthony (treasurer), Peabody Rifle Co. to Blacque Bey, Ottoman Minister to Washington, 10 July 1872; telegram, Saurent to Providence Tool Co., 27 July 1872.

29. RIHS, Providence Tool, Box 3, folder 1, telegram, Saurent, 27 July 1872; telegram, Blacque Bey, Ottoman Minister in Washington, to Yasu Pasha, Minister of War, 5 Aug. 1872.

30. RIHS, Providence Tool, Box 3, folder 1, telegram, Blacque Bey to John B. Anthony, 3 Sept. 1872; O. F. Winchester, president of Winchester Repeat-

ing Arms Co., to John B. Anthony, treasurer, Providence Tool Co., 17 Sept. 1872.

31. RIHS, Providence Tool, Box 3, folder 1, Winchester to Anthony, 24 Sept. 1872.

32. RIHS, Providence Tool, Box 3, folder 1, Winchester to Anthony, 4 Oct. 1872.

33. RIHS, Providence Tool, Box 3, folder 1, Winchester to Blacque Bey, 12 Oct. 1872.

34. RIHS, Providence Tool, Box 3, folder 2, Imperial Turkish Legation, Washington, D.C., to Anthony, 3 Jan. 1873; RIHS, Providence Tool, Box 3, folder 10, contract, 23 Aug. 1873.

35. WO 106/2, Col. Lennox to Constantinople, 13 Feb. 1877, no. 15.

36. RIHS, Box 3, folder 13, Robert L. Wheeler, "When Rhode Island Armed the Turks," *Providence Sunday Journal*, 16 Feb. 1941.

37. Deyrup, *Arms Makers of Connecticut Valley*, 210–212; Houze, *Winchester Repeating Arms*, 75, 97, 107; Ott, "Providence-Made Gun a Victim of Intrigue."

38. Bradley, *Guns for the Tsar*, 102–103.

39. Ibid., 105–109.

40. Ibid., 111, 116; L. G. Beskrovny, *The Russian Army and Fleet in the Nineteenth Century*, trans. Gordon E. Smith (Gulf Breeze, FL: Academic International Press, 1996), 176.

41. Bradley, *Guns for the Tsar*, 114–116.

42. WO 33/91, History of the Armstrong Gun, W. G. Armstrong, 14 July 1855. Report on the Construction of the Wrought-Iron Rifled Field Gun.

43. Krupp'sche Gussstahlfabrik, *Krupp 1812–1912* (Essen: Essen-Ruhr, 1912), 224.

44. Wilhelm Berdrow, *The Krupps: 150 Years Krupp History, 1787–1937*, trans. Fritz Homann (Berlin: Paul Schmidt, 1937), 169–171, 173–175.

45. *Krupp, a Great Business Man Seen through His Letters*, ed. Wilhelm Berdrow (New York: Dial Press, 1930), 264.

46. Ibid., 275.

47. Berdrow, *The Krupps*, 257.

48. Berdrow, ed., *Krupp, a Great Business Man*, 245.

49. Ibid., 351.

50. Ibid., 371.

51. Ibid., 368–369.

52. L. G. Beskrovnyi, *Russkaia armiia i flot v XIX veke* (Moscow: Nauka, 1973), 351; F. V. Greene, *Report on the Russian Army and the Campaigns in Turkey in 1877–78* (New York: D. Appleton and Co., 1879), 59.

53. KAE, WA 7f/886, Notiz Beziehungen zur Turkei; WO 106/1, Turkey, Con-

fidential 1869, 4; FO 78/2390, Elliot to Earl of Derby, 14 Nov. 1875; FO 78/2380, Elliot to Earl of Derby, 2 Feb. 1875.

54. FO 78/2390, Elliot to Earl of Derby, 14 Nov. 1875.

55. KAE, WA 4/1440, R. Cohnhein to C. Meyer, 10 Sept. 1878.

56. WO 106/2, Dangall to Layard, no. 9, 1878, 29–30.

57. WO 106/2, Lumley, British Legation to Brussels, to Earl of Derby, no. 98, 11 Aug. 1877, 17–18.

58. Thomas L. Kennedy, *The Arms of Kiangnan: Modernization in the Chinese Ordnance Industry, 1860–1895* (Boulder, CO: Westview Press, 1978), 35, 37, 40, 45, 48, 51, 59, 64; John L. Rawlinson, *China's Struggle for Naval Development, 1839–1895* (Cambridge, MA: Harvard University Press, 1967), 29, 32, 38.

59. WO 106/68, Major Mark S. Bell, Royal Engineer, Military Report on N. E. China, Mid-China, South China, vol. 2, 1884, 88–90.

60. Richard S. Horowitz, "Beyond the Marble Boat: The Transformation of the Chinese Military, 1850–1911," in David A. Graff and Robin Higham, eds., *A Military History of China* (Boulder, CO: Westview Press, 2002), 157.

61. ADM 231/2, China, Coast and River Defences and Government Arsenals, 84–87, 89–94.

62. WO 106/68, Bell, Military Report, 74–75.

63. Ibid., 76.

64. Kennedy, *Arms of Kiangnan,* 91–95.

65. WO 106/68, Bell, Military Report, 44, 49, 57–58; Udo Ratenhof, *Die Chinapolitik des Deutsches Reiches 1871 bis 1945* (Boppard am Rhien: Harald Boldt Verlag, 1987), 81; ADM 231/2, China, Coast and River Defences and Government Arsenals, 7; Zdenek Jindra, "Zur Entwickhing und Stellung der kanonenausfurh der Fifma Friedrich Krupp/Essen, 1854–1912," *Vierteljahrschrift für Sozial- und Wirtschafts-geschichte,* Beiheft 120 (Stuttgart: Franz Steiner Verlog, 1995), 967, 971–72.)

66. WO 106/68, Bell, Military Report, 77–82.

67. Kennedy, *Arms of Kiangnan,* 107–109.

68. WO 106/68, Bell, Military Report, 21; Rawlinson, *China's Struggle,* 62, 68–71, 140; Richard N. J. Wright, *The Chinese Steam Navy, 1862–1945* (London: Chatham, 2000), 42, 44, 46–47.

69. WO 106/68, Bell, Military Report, 15–16.

70. Wright, *Chinese Steam Navy,* 50–51, 53, 56; Rawlinson, *China's Struggle,* 77.

71. WO 106/68, Bell, Military Report, 20.

72. ADM 231/2, Chinese Navy, Sept. 1883; Rawlinson, *China's Struggle,* 110.

73. Rawlinson, *China's Struggle,* 80.

74. Ibid., 81.

75. WO 106/68, Bell, Military Report, 1–3.

76. ADM 231/2, China, Coast and River Defences and Government Arsenals, 84–87; 89–94.

77. Kennedy, *Arms of Kiangnan,* 150, 155.

78. John Scott Keltie, *Statesmen's Year-book,* 1879 (London: Palgrave), 662.

79. Richard J. Samuels, *"Rich Nation Strong Army": National Security and the Technological Transformation of Japan* (Ithaca, NY: Cornell University Press, 1994), 79.

80. Ibid., 81–82.

81. Chesneau and Kolesnik, eds., *Conway's All the World's Fighting Ships,* 219–220.

82. Marius B. Jansen, *The Making of Modern Japan* (Cambridge, MA: Belknap Press of Harvard University Press, 2000), 397.

83. Samuels, *"Rich Nation,"* 83–85.

84. Chesneau and Kolesnik, eds., *Conway's All the World's Fighting Ships,* 219–220.

85. Samuels, *"Rich Nation,"* 89.

86. Chesneau and Kolesnik, eds., *Conway's, All the Worlds Fighting Ships,* 216; David C. Evans and Mark A. Peattie, *Kaigun, Strategy Tactics and Technology in the Imperial Japanese Navy, 1887–1941* (Annapolis, MD: Naval Institute Press, 1997), 15.

87. Samuels, *"Rich Nation,"* 90.

88. Jansen, *Making of Modern Japan,* 376.

89. Statesman's Year-Book, 1879, 704.

90. Jansen, *Making of Modern Japan,* 421–422.

91. Stewart, "American Small Arms Industry," 61, 123–124, 131.

92. Ibid., 83, 123.

93. Ibid., 63, 126; Pam, *Royal Small Arms Factory,* 97.

94. Deyrup, *Arms Makers of Connecticut Valley,* 213.

2. Hand-Me-Down Guns

1. Mihail E. Ionescu, "The Military and the Establishment of the Romanian National State: Reciprocal Influences, 1856–1862," in *The Crucial Decade: East Central European Society and National Defense, 1859–1870,* ed. Bèla K. Kiràly (New York: Brooklyn College Press, 1984), 399.

2. Radu R. Florescu, "Cuza, Florescu, and Army Reform, 1859–1866," in Kiràly, ed., *The Crucial Decade,* 406; Gerald J. Bobango, "Foundations of the Independence Army: The Romanian Military," in Kiràly, ed., *The Crucial Decade,* 389; Florin Constantiniu, "Romania's Foreign Military Relations, 1859–1866," in Kiràly, ed., *The Crucial Decade,* 416.

3. FO 78/1514, Bulwer to Russell, 5 Dec. 1860.

4. Ibid., Bulwer to Earl Russell, 30 Dec. 1860; Dan Berindei, ed., *Documente*

Privind Domnia lui Alexandru Ioan Cuza, vol. 1 (1859–1861) (Bucarest: Editura Academiei Republicii Socialiste Romania, 1989), 74; FO 78/1569, Bulwer to Russell, 3 April 1861 (desp. 241).

5. FO 78/1514, Bulwer to Russell, 22 Dec. 1860, telegraphic; FO 78/1565, Bulwer to Russell, 16 Jan. 1861; FO 78/1566, Bulwer to Russell, 22 Jan. 1861; FO 78/1567, Bulwer to Russell, 27 Feb. 1861; FO 78/1569, Bulwer to Russell, 3 April 1861 (desp. 248); FO 78/1565, Bulwer to Russell, 9 Jan. 1861; FO 78/1569, Bulwer to Russell, 3 April 1861 (desp. 248).

6. FO 78/1569, Bulwer to Russell, 16 April 1861; FO 78/1649, Erskine to Russell, 4 March 1862; FO 78/1649, Longworth (consul-general) to Earl Russell, Belgrade, 1 Nov. 1862; FO 881/1182, Confidential Print. Correspondence Respecting the Importation of Arms into Serbia, during the Months of October, November, and December, 1862, Fane to Russell, 23 Oct. 1862; FO 881/1182, Ricketts to Longworth (Widdin), 23 Oct. 1862.

7. FO 881/1182, Consul-General Green to Russell, Bucharest, 22 Nov. 1862; ibid., Lord Bloomfield to Russell, Vienna, 20 Nov. 1862; ibid., Erskine to Russell, Constantinople, 24 Nov. 1862; ibid., Green to Russell, 26 Nov. 1862.

8. FO 881/1182, Green to Russell, 3 Dec. 1862.

9. Ibid., Russell to Green, 11 Dec. 1862.

10. Ibid., Russell to Green, 12 Dec. 1862.

11. Bobango, "Foundations of the Independence Army," 393; Constantiniu, "Romania's Foreign Military Relations," 417; Ilie Ceauşescu, "Romania's Military Policy and the National Liberation Struggle of the Peoples of Southeast and East Central Europe (1859–75)," in *War and Society in East Central Europe,* vol. 1, ed. Bèla K. Király and Gunther E. Rothenberg (New York: Brooklyn College Press, 1975), 257–258.

12. Ioan Talpeş, "Romanian Military Collaboration with East Central European Struggles for Emancipation and National Unity, 1856–1866," in Király, ed., *The Crucial Decade,* 423.

13. FO 881/1182, Russell (London) to Longworth, 9 Dec. 1862; Lord Napier to Earl Russell, 17 Dec. 1862.

14. FO 881/1182, Acting Consul-General Wassitsch, 28 Dec. 1862.

15. FO 881/1151, Major Edward Gordon, R. E., Military Report on Serbia, 19 June 1863; Gale Stokes, "Serbian Military Doctrine and the Crisis of 1875–78," in *Insurrections, Wars, and the Eastern Crisis in the 1870s,* ed. Bèla K. Király and Gale Stokes (Boulder, CO: Social Science Monographs, 1985), 265–267; Milard E. Kmečić, "The Serbian Army in the Wars of 1876–78: National Liability or National Asset?" in Király and Stokes, eds., *Insurrections,* 282–283.

16. FO 881/1182, A. Buchanan to Russell, 20 Dec. 1862.; ibid., Lord Napier to Earl Russell, St. Petersburgh, 11 Dec. 1862.

17. Ibid., Russell to Longworth, 11 Dec. 1862.
18. Ibid., Russell to Longworth, 12 Dec. 1862.
19. Ibid., Bulwer to Russell, 22 Jan. 1863; ibid., Lord Napier to Earl Russell, St. Petersburgh, 11 Dec. 1862; ibid., Lord Russell to Lord Napier, 17 Dec. 1862.
20. Ibid., Consul-General Murray to Earl Russell, Odessa, 9 March 1863; ibid., Napier to Russell, 21 April 1863.
21. Florescu, "Cuza," 406; Ceaușescu, "Romania's Military Policy," 261; Bobango, "Foundations of the Independence Army," 393; Ionescu, "The Military and the Establishment of the Romanian National State," 399.
22. Bobango, "Foundations of the Independence Army," 393.
23. Dan Berindei, "The Romanian Armed Forces in the Eighteenth and Nineteenth Centuries," in Király and Rothenberg, eds., *War and Society,* 237.
24. FO 78/2022, Elliot to Lord Stanley, 5 Sept. 1868.
25. Ibid.
26. FO 78/2023, Elliot to Lord Stanley, 24 Oct. 1868.
27. FO 78/3097, Goschen to Granville, 31 Oct. 1880; FO 104/28, White to Granville, 23 Dec. 1882; FO 78/3753, White to Salisbury, 29 Sept. 1885, enclosure.
28. RGVIA f. 430, op. 1, d. 165, ch. 1, ll. 6–8, 16, 34, 39, 47, 90.
29. Harold G. Marcus, *A History of Ethiopia* (Berkeley: University of California Press, 1994), 68; Richard Pankhurst, *Economic History of Ethiopia, 1800–1935* (Addis Ababa: Haile Sellassii I University Press, 1968), 583; Pankhurst, *Economic History,* 585; Sven Rubenson, ed., *Tewodros and His Contemporaries, 1855–1868,* Acta Aethiopia vol. 2 (Addis Ababa: Addis Ababa University Press, 1994), 101; *BDFA,* vol. 13, 48.
30. Marcus, *History of Ethiopia,* 70; Pankhurst, *Economic History,* 585.
31. Rubenson, ed., *Tewodros,* 298.
32. *BDFA,* vol. 13, 69–71.
33. Pankhurst, *Economic History,* 587; Marcus, *History of Ethiopia,* 72.
34. Pankhurst, *Economic History,* 587–588; Marcus, *History of Ethiopia,* 73–75; *BDFA,* vol. 13, 76–77.
35. Marcus, *History of Ethiopia,* 75; Pankhurst, *Economic History,* 589.
36. Pankhurst, *Economic History,* 590; Marcus, *History of Ethiopia,* 78.
37. Richard Caulk, *"Between the Jaws of Hyenas": A Diplomatic History of Ethiopia (1876–1896)* (Wiesbaden: Harrassowitz Verlag, 2002), 24–25, 28; Pankhurst, *Economic History,* 590, 591; Caulk, *"Between the Jaws of Hyenas,"* 28.
38. Caulk, *"Between the Jaws of Hyenas,"* 25.
39. FO 403/82, Baring to Granville, 8 Dec. 1884; ibid., Granville to J. S. Lumley, 29 Dec. 1884.

40. FO 403/83, Lumley to Granville, 5 Jan. 1885; ibid., Lumley to Granville, 13 Jan. 1885.

41. FO 403/84, King John of Ethiopia to Queen Victoria, 10 Nov. 1885.

42. FO 403/91, King John to Queen Victoria, 19 Oct. 1887.

43. Pankhurst, *Economic History,* 592–594; Marcus, *History of Ethiopia,* 83–84.

44. Caulk, *"Between the Jaws of Hyenas,"* 69; Pankhurst, *Economic History,* 594.

45. *BDFA,* vol. 13, 107.

46. Ibid., 107–108.

47. FO 403/91, Consul Stace to E. Barring, 18 Dec. 1887.

48. Ibid., E. V. Stace, Report, 4 Jan. 1888.

49. Caulk, *"Between the Jaws of Hyenas,"* 96, 99; Marcus, *History of Ethiopia,* 85, 87; Pankhurst, *Economic History,* 596.

50. FO 403/91, Salisbury to Lytton, 17 Feb. 1888.

51. Pankhurst, *Economic History,* 595, 597.

52. FO 403/91, Lytton to Salisbury, 28 March 1888.

53. FO 403/91, Godley to Foreign Office, 16 April 1888; ibid., Lytton to Salisbury, 25 April 1888; ibid., Military Attaché, Rome, to Salisbury, 30 April 1888.

54. Ibid., Lytton to Salisbury, 28 April 1888; ibid., Lytton to Salisbury, 18 May 1888; ibid., Salisbury to Lytton, 6 June 1888.

55. FO 403/123, Lytton to Salisbury, 2 Feb. 1889; ibid., Baring to Salisbury, 29 Jan. 1889; ibid., Catalani to Foreign Office, 18 Feb. 1889.

56. FO 403/124, Stace to Baring, 10 Nov. 1889.

57. Ibid., Walsh to Stace, 25 May 1889.

58. Bairu Tafla, ed., *Ethiopian Records of the Menilek Era: Selected Amharic Documents from the Nachlass of Alfred Ilg, 1884–1900* (Wiesbaden: Harrassowitz Verlag, 2000), 363–364.

59. Ibid., 379, 381.

60. Ibid., 383.

61. Ibid., 384.

62. Ibid., 386–387.

63. Ibid., 367, 369.

64. FO 403/84, Gissing to Stace, 6 Sept. 1888, enclosed in Portal to Salisbury, 2 Oct. 1888; ibid., Baring to Salisbury, 21 Nov. 1888.

65. Marcus, *History of Ethiopia,* 89–90; FO 403/124, Dufferin to Salisbury, 28 Nov. 1889; ibid., India Office to Foreign Office, 19 Dec. 1889.

66. Caulk, *"Between the Jaws of Hyenas,"* 179, 226, 240.

67. FO 403/126, Salisbury to Baring, 10 Nov. 1890; ibid., Portal to Salisbury, 25 Sept. 1890; ibid., Stace to Baring, 29 May, 1890.

68. FO 403/155, Stace to Hardinge, 2 Sept. 1891.

69. Tafla, ed., *Ethiopian Records*, 410.
70. FO 403/155, Salisbury to Dufferin, 4 Nov. 1891.
71. FO 403/190, Ras Makunan to Stace, 8 Nov. 1893.
72. FO 403/221, India Office to Foreign Office, 8 Feb. 1895; ibid., Jopp to Governor of Bombay, 3 Jan. 1895; ibid., India Office to Foreign Office, 18 March 1895.
73. FO 403/155, India Office to Foreign Office, 4 Nov. 1891.
74. Ibid., Foreign Office to India Office, 11 Nov. 1891; FO 403/190, India Office to Foreign Office, 26 July 1893.
75. FO 403/190, Vivian to Rosebery, 24 May 1893; FO 403/221, Clare Ford to Kimberley, 4 Jan. 1895; *BDFA*, vol. 13, 148.
76. Caulk, "*Between the Jaws of Hyenas,*" 375, 397; FO 403/206, Count Tornielli to Foreign Office, 7 Aug. 1894.
77. FO 403/190, Vivain to Rosebery, 2 June 1893.
78. FO 403/221, Secret Report to Italian Ministry for Foreign Affairs, 9 Jan. 1895.
79. Ibid., Clare Ford to Kimberley, 8 April 1895; ibid., Kimberley to Clare Ford, 10 April 1895.
80. Ibid., Kimberley to Clare Ford, 9 April 1895.
81. Tafla, ed., *Ethiopian Records*, 440–441.
82. Ibid., 441–442.
83. FO 403/221, W. B. Ferris, Memorandum, 4 April 1895.
84. Ibid., Ras Makunan to Consul, Somali Coast, 10 June 1895.
85. Ibid., Ferris to Cromer, 15 July 1895.
86. FO 403/255, Foreign Office to India Office, 30 Jan. 1897; ibid., Cuningham to Hamilton, 6 April 1897.
87. FO 403/221, Clare Ford to Kimberley, 5 Jan. 1895.
88. Tafla, ed., *Ethiopian Records*, 436.
89. FO 403/221, Rumbold to Salisbury, 20 July 1896; ibid., Rumbold to Salisbury, 10 Aug. 1896; Caulk, "*Between the Jaws of Hyenas,*" 603–604, 647.
90. FO 403/221, F. Plunkett to Salisbury, 15 Dec. 1895; ibid., Secret messenger sent to Jibuti and Obokh, 1 Jan. 1896; ibid., Signor Adamoli to General Ferrera, 10 Feb. 1896; ibid., enclosure in no. 57, *Djibouti News*, 25 Feb. 1896.
91. Ibid., Memorandum communicated by Italian Ambassador, 17 Aug. 1896; FO 403/274, Plunkett (Brussels) to Salisbury, 15 Jan. 1898.

3. Arms Trade Colonialism

1. FO 403/274, Manson (Paris) to Salisbury, 14 Jan. 1898; FO 403/298, Intelligence Division to Foreign Office, 12 April 1900.
2. FO 403/376, Capt. Cordeaux to Elgin, 10 April 1906.

3. Ibid.

4. Ibid., Count de Bosdari to Grey, 28 May 1906; ibid., Signor Tittoni to des Graz, 24 July 1906; ibid., Count de Bosdari to Grey, 8 Aug. 1906; ibid., Hardinge to Grey, 28 Sept. 1906.

5. Ibid., Extracts from Dispatch from Commissioner Swayne to Secretary of State, 11 Jan. 1906; ibid., War Office to Foreign Office, 28 Nov. 1906, enclosure 1.

6. FO 403/376, Agreement France, Britain, and Italy, 13 Dec. 1906; Richard Pankhurst, *Economic History of Ethiopia, 1800–1935* (Addis Ababa: Haile Sellassii I University Press, 1968), 605.

7. FO 428/1, Foreign Office to Colonial Office, 18 Jan. 1907; ibid., Hohler to Grey, 18 Oct. 1907; ibid., Hohler to Grey, 22 May 1907, enclosure.

8. Ibid., Hohler to Grey, 4 July 1907.

9. Ibid., Hohler to Grey, 3 Sept. 1907.

10. Ibid., San Giuliano to Grey, 13 Aug. 1907; ibid., Hohler to Grey, 30 Oct. 1907.

11. Ibid., Gerolimato to G. Clerk, 12 Dec. 1907.

12. FO 428/2, G. Colli, Report Addressed to Italian Foreign Minister, 28 Dec. 1907.

13. FO 428/1, Hohler to Grey, 30 May, 1907.

14. Ibid., Hohler to Grey, 1 Aug. 1907.

15. Ibid.

16. FO 428/2, Harrington to Grey, 21 March 1908.

17. FO 428/5, Herbert Hervey (Addis) to Grey, 23 Oct. 1909; FO 401/14, Thesiger to Grey, 14 May 1910.

18. FO 428/13, Thesiger to Grey, 13 Aug. 1913.

19. FO 428/2, London Chamber of Commerce to Grey, 15 April 1908.

20. Ibid., Hornby to Grey, 16 April 1908.

21. FO 428/3, H. F. Ward to G. R. Clerk (Harrar), 30 June 1908.

22. Ibid., Ward to Churchill (Abyssinia), 5 July 1908.

23. Ibid., Bertie to Grey, 8 Aug. 1908.

24. FO 428/5, India Office to Foreign Office, June 1909; ibid., Foreign Office to Colonial Office, 28 Sept. 1909.

25. Ibid., Bertie to Grey, 8 Oct. 1909; ibid., Grey to Bertie, 18 Oct. 1909.

26. FO 428/6, Signor Martin-Franklin to Grey, 10 May 1910; *BDFA,* vol. 13, 338–339.

27. FO 428/7, MacDonald (Tokyo) to Grey, 2 Dec. 1910; FO 428/8, Wilfred Thesiger to Grey, 27 Dec. 1910; ibid., MacDonald to Grey, 16 March 1911; ibid., Rumbold to Grey, 11 May 1911; ibid., Rumbold to Grey, 12 May 1911; ibid., Rumbold to Grey, 3 June 1911; FO 428/9, MacDonald to Grey, 28 Oct. 1911.

28. FO 428/8, Wilfred Thesiger to Grey, 27 Dec. 1910; ibid., Thesiger to Grey, 12 Jan. 1911; FO 428/10, Thesiger to Grey, 14 Feb. 1912; ibid., Doughty-Wylie to Grey, 12 Dec. 1911.

29. FO 428/10, Thesiger to Grey, 14 Feb. 1912.

30. Ibid., Doughty-Wylie to Grey, 12 Dec. 1911.

31. Ibid., Thesiger to Grey, 14 Feb. 1912.

32. FO 428/11, Director of Military Operations to Foreign Office, 20 July 1912; ibid., Foreign Office to Colonial Office, 1 Aug. 1912; ibid., Thesiger to Grey, 27 June 1912.

33. Ibid., Bertie to Grey, 13 Sept. 1912; ibid., Thesiger to Grey, 12 Sept. 1912.

34. *BDFA,* vol. 13, 392–393, 408.

35. *BDFA,* vol. 13, 392–393; FO 428/13, Thesiger to Grey, 13 Aug. 1913; FO 428/12, Grey to Bertie, 17 March 1913; ibid., Thesiger to Grey, 14 March 1913; ibid., Thesiger to Grey, 6 May 1913; ibid. Thesiger to Grey, 2 June 1913.

36. FO 428/13, Thesiger to Grey, 15 Aug. 1913; *BDFA,* vol. 13, 403, 408.

37. *BDFA,* vol. 13, 409.

4. Austro-German Hegemony in Eastern Europe

1. Robert W. D. Ball, *Mauser Military Rifles of the World* (Iola, WI: Krause, 1996), 8, 204; Wolfgang Seel, *Mauser, von der Waffenschmiede zum Weltunternehmen* (Zürich: Verlag Stocker-Schmid AG, 1988), 27–30.

2. Michaela Pfaffenwimmer, "Die wirtschaftliche und soziale Entwicklung der 'Österreichischen Waffenfabriks-Aktiengesellschaft' unter der Leitung des Generaldirektors Josef Werndl, 1869–1889" (Ph.D. dissertation, Universität Wien, 1985), 165–166; Martin Gutsjahr, "Rüstungsunternehmen Österreich-Ungarns vor und im Ersten Weltkrieg: Die Entwicklung dargestellt an die Firmen Skoda, Steyr, Austro-Daimler und Lohner" (Ph.D. dissertation, Universität Wien, 1995), 5, 59–60, 90–97; Manfried Reinschedl, "Die Rüstung Österreich-Ungarns von 1880 bis zum Ausbruch des Ersten Weltkriegs" (Master's thesis, Universität Wien, 1996), 101–105; *125 Jahre Waffen aus Steyr* (Steyr: Steyr Mannlicher Ges. M.B.H., 1989), 6, 9.

3. ADM 231/10, Report 119, Shipbuilding Policy of Foreign Nations, 1886.

4. Justin McCarthy, *The Ottoman Turks* (New York: Addison Wesley Longman, 1997), 304–313.

5. Baron Wladimir Giesl v. Gieslingen, *Zwei Jahrzehnte im nahen Orient* (Berlin: Verlag für Kulturpolitik, 1927), 30.

6. Gieslingen, *Zwei,* 49; William Hale, *Turkish Politics and the Military* (London: Routledge, 1994) 28–29; Eric Zürcher, *Turkey, a Modern History* (London: I. B. Taurus, 1997), 84.

7. Jehuda Wallach, *Anatomie einer Militarhilfe, Die preussich-deutschen Militarmissionen in der Turkei, 1835–1914* (Dusseldorf: Droste Verlag, 1976), 35, 43, 54, 64, 85.

8. Gieslingen, *Zwei,* 46

9. FO 78/4479, Chermside to Ford, 26 May 1893, enclosed in Ford to Rosebery, 29 May 1893.

10. Sir Henry F. Woods, *Spunyarn from the Strands of a Sailor's Life Afloat and Ashore,* vol. 2 (London: Hutchinson, 1924), 147.

11. Hale, *Turkish Politics,* 29.

12. For a thorough discussion of the German military mission, see Wallach, *Anatomie,* 35–85.

13. FO 78/4479, Ford to Earl of Rosebery, 24 April 1893; ibid., Ford to Rosebery, 26 May 1893; enclosure, Chermside to Ford, Position of German Military Mission at Constantinople.

14. FO 78/3869, White to Rosebery, 11 Feb. 1886; FO 78/4098, White to Salisbury, 18 Feb. 1888; Ilber Ortayli, *Ikinci Abdulhamit Döneminde Osmanli Imparatorlugunda Alman Nüfuzu* (Ankara: Ankara Üniversitesi Siyasi Biligiler Fakultesi Yayinlari no. 479, 1981), 65–68; Wallach, *Anatomie,* 105; Generalfeldmarschall Colmar Freiherr von der Goltz, *Denkwürdigkeiten* (Berlin: Mittler & Sohn, 1929), 124–125, 140.

15. Ball, *Mauser Military,* 236–239; Seel, *Mauser,* 33–34; *Geschichte der Mauser-Werke* (Berlin: VDI-Verlag, 1938), 89–94.

16. FO 78/4002, White to Foreign Office (telegram), 30 Jan. 1887.

17. FO 78/4001, White to Salisbury, 28 Nov. 1887.

18. Ball, *Mauser Military,* 236–239; Seel, *Mauser,* 33–34; *Geschichte der Mauser-Werke,* 89–94; FO 78/4342, Chermside to White, 16 Jan. 1891.

19. Wallach, *Anatomie,* 105; Rafii-Sukru Suvla, "The Ottoman Debt, 1850–1939," in Charles Issawi, ed., *The Economic History of the Middle East, 1800–1914* (Chicago: University of Chicago Press, 1975), 104; Justin McCarthy, *The Arab World, Turkey, and the Balkans: A Handbook of Historical Statistics* (Boston: G. K. Hall, 1982), 160, 184; FO 78/4276, Chermside to Fane, 5 July 1890; FO78/4479, Ford to Rosebery, 24 April 1893; Lothar Rathman, *Berlin-Bagdad* (Berlin: Dietz Verlag, 1962), 18; Wallach, *Anatomie,* 105; Ortayli, *Ikinci,* 68; FO 78/4479, Ford to Earl of Rosebery, 24 April 1893.

20. KAE, 1891, FAH 3C/217, Krupp, 21 Dec. 1891.

21. Gieslingen, *Zwei,* 34, 21, 29.

22. SA, SS 0225, Dossier concernent les négociations et les conventions entre MM. Schneider et Cie et divers pays étrangers désireux d'acquérir du materiel naval, no. 7, 22 Jan. 1894.

23. SA, SS 0225, no. 14, 18 May 1894; no. 18, 28 July 1894; no. 26, 8 Jan. 1895; no. 32, 25 June 1895; no. 33, 1 Oct. 1895.

24. SA, SS 0225, no. 36, 9 March 1896.
25. SA, SS 0225, no. 37, 25 March 1896.
26. KA, Generalstab. Militarattaches. Konstantinopel, kt. 39, Res. 46, 28 March 1899.
27. Wallach, *Anatomie,* 77, 105; Ortayli, *Ikinci,* 68.
28. FO 78/4706, Currie to Salisbury, 19 April 1896, enclosure 1.
29. FO 78/4105, White to Salisbury, 6 Nov. 1888; ibid., White to Salisbury, 15 Nov. 1888, enclosure, memorandum, 14 Nov. 1888.
30. FO 78/4106, White to Salisbury, 7 Dec. 1888.
31. RGVIA f. 450, op. 1, d. 118, l.62.
32. ADM 231/31, Capt. Hugh Williams, RN, Report 580, Turkish Fleet and Dockyards, etc. 1900.
33. FO 78/3751, White to Granville, 29 April 1885; ADM 231/10, Capt. Henry Kane, Report 127, Turkish Fleet and Dockyards, 1886.
34. FO 78/4276, Chermside to Fane, 5 July 1890; ADM 231/10, Report 127, Turkish Fleet and Dockyards, 1886.
35. ADM 231/10, Report 127, Turkish Fleet and Dockyards, 1886.
36. Bernd Langensiepen and Ahmet Güleryüz, *The Ottoman Steam Navy, 1828–1923,* ed. and trans. James Cooper (Annapolis, MD: Naval Institute Press, 1995), 9–10; KAE, FAH 3C/205, Friedrich Krupp to Kontre-Admiral Frieherrn v. Senden-Birbran, 9 Nov. 1897.
37. Andrea Filippo Saba, "'L'attività dell'Ansaldo nell'Impero Ottomano," in Peter Hertner, ed., *Storia dell'Ansaldo 3. Dai Bombrini ai Perrone, 1903–1914* (Bari: Gius. Laterza & Figli, 1996), 95–99; Langensiepen and Güleryüz, *Ottoman Steam,* 10–11.
38. Langensiepen and Güleryüz, *Ottoman Steam,* 11; TWAS 130/1266, Armstrong Minute Books, Meetings of Directors, 9 June 1898, 22 May 1901, 18 Sept. 1901, 29 Oct. 1902, 26 Nov. 1902; TWAS 130/1267, Armstrong Minute Books, Meeting of Directors, 19 May 1904, 20 July 1904.
39. Joseph Bradley, *Guns for the Tsar: American Technology and the Small Arms Industry in Nineteenth-Century Russia* (Dekalb: Northern Illinois University Press, 1990) 170; L. G. Beskrovny, *The Russian Army and Fleet in the Nineteenth Century,* trans. Gordon E. Smith (Gulf Breeze, FL: Academic International Press, 1996), 181.
40. RGVIA f. 514, op. 1, d. 52, ll. 2–7.
41. FO 65/1700, Napier to Spring Rice, 24 May 1905.
42. Ibid.
43. *50 Jahre Rheinmetall Düsseldorf, 1889–1939* (Düsseldorf: Rheinmetall-Borsig Aktiengesellschaft, 1939), 32.
44. Eleanor C. Barnes, *Alfred Yarrow: His Life and Work* (London: Edward Arnold, 1924), 82–85; N. Grechaniuk, V. Dmitriev, F. Krinitsyn, Iu. Cher-

nov, *Baltiiskii flot* (Moscow: Voennoe Izd. Ministerstva Oborony Soiuz SSR, 1960), 102.

45. Barnes, *Alfred Yarrow*, 95, 105–106.

46. ADM 231/20, Report 293, Russia. Guns, Torpedoes, Fortifications of Cronstadt, 1891; ADM 231/23, Report 364, Russia. Coast Defence Ordnance and Torpedo Material, 1894; ADM 231/31, Report 567, Russia, Fleet, Dockyards, 1900; ADM 231/38, Report 700, Foreign Naval Progress and Estimates, 1903.

47. ADM 231/23, Report 364, Russia. Coast Defence Ordnance and Torpedo Material, 1894.

48. L. G. Beskrovnyi, *Russkaia armiia i flot v XIX veke* (Moskva: Nauka, 1973), 529; ADM 231/33, Report 602, Lists Showing the Capabilities of France, Russia, Germany, Italy, United States for Building and Completing Vessels of War, 1900; ADM 231/48, Report 820, Shipbuilding Capabilities of the Principal Naval Powers, Feb. 1907; ADM 231/34, Report 624, Germany, Harbours and Dockyards, 1901; ibid., Report 628. Reports on Foreign Naval Affairs, 1901; FO 65/1553, O'Conor to Salisbury, 29 March 1898; FO 65/1564, Mackie to Salisbury, 1 Aug. 1898.

49. FO 65/1578, Scott to Salisbury, 22 May 1899.

50. FO 104/44, Sanderson to Salisbury, 18 Oct. 1885; FO 78/3894, Stephen to Iddesleigh, 12 Aug. 1886; FO 78/4137, Hardinge to Salisbury, 9 Aug. 1888; ibid., O'Conor to Salisbury, 9 July 1888; FO 78/3894, O'Conor to Salisbury, 15 Oct. 1889.

51. FO 78/3894, O'Conor to Salisbury, 15 Oct. 1889; ibid., O'Conor to Salisbury, 23 Nov. 1889; FO 78/4377, R. W. Graves to O'Conor, 7 May 1891.

52. FO 78/4377, R. W. Graves to O'Conor, 7 May 1891; FO 881/6253, Notes on the Bulgarian Army by Col. Chermside, London, War Office, 1892; Raymond Poidevin, *Les Relations Économiques et Financières entre La France et L'Allemagne de 1898 a 1914* (Paris: Armand Colin, 1969), 147.

53. FO 881/6253, Notes on the Bulgarian Army by Col. Chermside, London, War Office, 1892.

54. Ibid.

55. FO 78/4753, Elliot to Salisbury, 14 Feb. 1896, enclosure, speech by Col. Petrov, Minister of War to Sobranie.

56. SA, SS 0225, no. 8, 27 Jan. 1894; ibid., no. 18, 28 July 1894; SA, FX 0043–06, Mission de M. Michel-Schmidt a Sofia.

57. Poidevin, *Les Relations,* 55–57; FO 78/5146, Elliot to Lansdowne, 2 May 1901; FO 78/4853, Elliot to Salisbury, 22 Feb. 1897; ibid., Elliot to Salisbury, 25 March 1897.

58. FO 78/4951, Elliot to Salisbury, 5 July 1898.

59. FO 78/4854, Elliot to Salisbury, 19 May 1897; ibid., Elliot to Salisbury, 29 May 1897.

60. FO 78/4542, Chermside to Philip Currie, 11 July 1894; FO 78/4853, Elliot to Salisbury, 6 Jan. 1897; ibid., Elliot to Salisbury, 25 Jan. 1897; FO 78/4854, Elliot to Salisbury, 20 May 1897; ibid., Elliot to Salisbury, 2 June 1897.

61. RGVIA f. 439, op. 1, d. 18, l. 26; MAE, Serbie 1, dossier 8. I am indebted to Mike Creswell for providing me with these materials. FO 105/19, Locock to Granville, 1 June 1881; Ball, *Mauser Military,* 8, 204; Seel, *Mauser,* 27–30.

62. FO 105/51, Wyndham to Salisbury, 11 Dec. 1885; ibid., Wyndham to Salisbury, 16 Dec. 1885; FO 105/55, Wyndham to Salisbury, 21 Jan. 1886; FO 105/56, Wyndham to Rosebery, 11 March 1886; FO 78/3894, Stephen to Earl of Iddesleigh, 12 Aug. 1886.

63. FO 105/51, Wyndham to Salisbury, 4 Nov. 1885; FO 105/63, Wyndham to Salisbury, 20 Feb. 1887; ibid., Wyndham to Salisbury, 16 April 1887.

64. FO 881/6017, Capt. C. E. Callwell, RA, Armed Strength of Serbia, 1890; FO 105/90, St. John to Salisbury, 12 Oct. 1891; O. V. Pavliuchenko, *Rossiia i Serbiia, 1888–1903* (Kiev: Naukova Dumka, 1987), 16.

65. RGVIA f. 439, op. 1, d. 21, ch. 1, ll. 1, 5, 7–8, 15, 19; FO 105/95, St. John to Salisbury, 27 Feb. 1892.

66. RGVIA f. 439, op. 1, d. 21, ch. 1, ll. 19, 47, 49.

67. RGVIA f. 439, op. 1, d. 21, ch. 1, l. 62; Poidevin, *Les Relations,* 61.

68. RGVIA f. 439, op. 1, d. 21, ch. 1, l. 108, 146.

69. KA, Generalstab. Militarattaches. Belgrad, kt. 4, Res. 85, 30 June 1899; FO 105/117, Elliot to Salisbury, 23 July 1897; FO 105/118, Elliot to Salisbury, 3 Aug. 1897; FO 105/127, Goschen to Salisbury, 22 Oct. 1899.

70. FO 105/127, Goschen to Salisbury, 22 Oct. 1899; WO 106/6177, Reports on Changes in Various Foreign Armies during the Year 1900; FO 105/132, Macdonald to Salisbury, 20 Feb. 1900; FO 105/133, Macdonald to Salisbury, 7 Sept. 1900; ibid., Bonham to Lansdowne, 21 Nov. 1900; WO 106/6184, Report on Changes in Foreign Armies during 1907.

71. FO 105/45, Locock to Granville, 3 March 1884; FO 105/50, Wyndham to Granville, 9 June 1885; FO 105/51, Wyndham to Salisbury, 5 Dec. 1885.

72. FO 105/89, St. John to Salisbury, 3 Feb. 1891; FO 105/90, Lyon to Salisbury, 28 July 1891.

73. FO 105/95, St. John to Salisbury, 5 Feb. 1892.

74. Ibid., St. John to Salisbury, 25 Feb. 1892; ibid., St. John to Salisbury, 27 Feb. 1892.

75. Ibid., St. John to Salisbury, 19 April 1892.

76. KA, Generalstab. Militarattaches. Belgrad, kt. 4, Res. 90, 3 July 1899; FO 105/117, Fane to Salisbury, 9 June 1897; KA, Generalstab. Militarattaches. Belgrad, kt 5. Res. 8, 23 Jan. 1901.
77. FO 105/133, Bonham to Lansdowne, 13 Dec. 1900.
78. KA, Generalstb. Militarattaches. Belgrad, kt. 4, Sekret Res. 22, 15 March 1900.
79. FO 105/143, Bonham to Lansdowne, April 21, 1902.
80. HHSA, Konsulatsarchiv Belgrad, kt. 11, Polit. Und Admin, Jahres-Bericht des k.u.k. Konsulats in Belgrad pro 1902, 81–84.
81. Charles and Barbara Jelavich, *The Establishment of the Balkan National States, 1804–1920* (Seattle: University of Washington Press, 1977), 171–175.
82. FO 32/519, Corbett to Granville, 1 May 1880; ibid., Corbett to Granville, 14 July 1880; FO 32/520, Corbett to Granville, 14 Dec. 1880.
83. FO 32/551, Ford to Granville, 30 Aug. 1883.
84. Ibid.
85. ADM 231/3, Greece. Report on Dockyards and Coast Defenses. Admiralty, Foreign Intelligence Committee (no. 18) October 1883; FO 32/557, Nicolson to Granville, 20 Nov. 1884.
86. FO 32/520, Corbett to Granville, 7 Dec. 1880; ibid., Corbett to Granville, 13 Oct. 1880; FO 32/565, Rumbold to Salisbury, 10 Dec. 1885; FO 32/588, Carew to Salisbury, 4 June 1887; FO 32/609, Monson to Salisbury, 29 Nov. 1889.
87. FO 32/729, Egerton to Salisbury, 9 July 1901.
88. FO 32/729, Egerton to Salisbury, note, 16 July 1901.
89. FO 32/520, Corbett to Granville, 18 July 1880; ibid., Corbett to Granville, 7 Aug. 1880; FO 32/528, Corbett to Granville, 26 Feb. 1881; FO 421/68, Col. Fraser to Paget, 11 Jan. 1886, Confidential Print. Affairs of South-Eastern Europe. January 1886; ADM 231/23 Report 370, Greece. Coasts Defenses, etc. 1893; FO 32/677, Egerton to Salisbury, 6 Jan. 1896.
90. FO 104/28, White to Granville, 24 July 1882; ibid., White to Granville, 6 Dec. 1882; ibid., White to Granville, 13 March 1882.
91. FO 104/33, White to Granville, 14 Nov. 1883.
92. FO 104/70, Lascelles to Salisbury, 27 Feb. 1888.
93. Ibid., Lascelles to Salisbury, 16 April 1888.
94. FO 104/44, Sanderson to Salisbury, 13 Oct. 1885.
95. *Krupp, a Great Business Man Seen through His Letters,* ed. Wilhelm Berdrow (New York: Dial Press, 1930), 409.
96. FO 104/126, Wyndham to Salisbury, 5 May 1896; RGVIA f. 2000, op. 1, d. 794, l. 1; RGVIA f. 2000, op. 1, d. 796, l. 2; FO 104/144, Kennedy to Lansdowne, 5 Dec. 1900; ibid., Kennedy to Lansdowne, 18 Dec. 1900; RGVIA f. 2000, op. 1, d. 794, l. 2.

97. RGVIA f. 2000, op. 1, d. 794, ll. 8, 11; FO 104/151, Kennedy to Lansdowne, 24 Jan. 1902; ibid., Kennedy to Lansdowne, 11 March 1902; ibid., Kennedy to Lansdowne, 8 April 1902.

98. FO 104/151, Kennedy to Lansdowne, 16 April 1902; Poidevin, *Les Relations,* 59, 306, 309.

99. FO 104/151, Kennedy to Lansdowne, 15 May, 1902.

100. RGVIA f. 2000, op. 1, d. 796, l.8; Poidevin, *Les Relations,* 310.

101. Poidevin, *Les Relations,* 341.

5. A Tale of Two Arms Races

1. David Rock, *State Building and Political Movements in Argentina, 1860–1916* (Stanford, CA: Stanford University Press, 2002), 38, 53; Frederick M. Nunn, "The South American Military Tradition: Preprofessional Armies in Argentina, Chile, Peru, and Brazil," in Linda Alexander Rodriguez, ed., *Rank and Privilege: The Military and Society in Latin America* (Wilmington, DE: Scholarly Resources, 1994) 73, 76.

2. George v. Rauch, *Conflict in the Southern Cone: The Argentine Military and the Boundary Dispute with Chile, 1870–1902* (Westport, CT: Praeger, 1999), 115; FO 13/337, Foreign Office to Council Office, 10 Aug. 1855; FO 13/369, C. Moreira, Brazilian Legation, to Clarendon, 27 Jan. 1856; FO 13/347, C. Moreira, Brazilian Legation, to Clarendon, 17 Jan. 1856; Rock, *State Building,* 41.

3. FO 6/324, Argentine Legation to Granville, 5 Feb. 1874; FO 6/490, Haggard (Buenos Aires) to Lansdowne, 13 May 1905, enclosure, Juan A. Martin, "Memoria del Departamento de Marina, 1904–1905"; VHS, Correspondence of Thomas J. Page, Mss 1 P1465b26–51; Rauch, *Southern Cone,* 115, 119.

4. FO 6/344, Admiralty to Foreign Office, 10 May 1877; FO 6/351, Admiralty to Foreign Office, 8 Oct. 1878; FO 6/366, Argentine Legation to Granville, 17 March 1881.

5. FO 6/364, Rumbold to Granville, 24 Feb. 1881.

6. FO 6/366, Cmdr. Crowley, Confidential Report on Argentine Military and Naval Forces, 18 Feb. 1881.

7. FO 6/490, Haggard (Buenos Aires) to Lansdowne, 13 May 1905, enclosure, Juan A. Martin, "Memoria del Departamento de Marina, 1904–1905."

8. FO 6/366, Report from Capt. Loftus Jones, enclosed in Hall to Under Secretary of State, 10 May 1881.

9. Ibid.; FO 13/587, Report of Brazilian Minister of Marine, Sabina Elay Pessoa, 1882, enclosure in Rio Consulate to Granville, 12 April 1883; FO 13/588, Corbett to Granville, 18 July 1883; ADM 231/4, Brazil, Reports, Admiralty Intelligence Committee, March 1884.

10. Rauch, *Southern Cone,* 119, 126n.42; FO 6/490, Haggard (Buenos Aires) to Lansdowne, 13 May 1905, enclosure, Juan A. Martin, "Memoria del Departamento de Marina, 1904–1905."

11. Bruce W. Farcau, *The Ten Cents War: Chile, Peru, and Bolivia in the War of the Pacific, 1879–1884* (Westport, CT: Prager, 2000), 16–17; Robert L. Scheina, *Latin America's Wars,* vol. 1, *The Age of the Caudillo, 1791–1899* (Washington, DC: Brassey's, 2003), 333–334.

12. *BDFA,* part I, Series D, Latin America, 1845–1914, vol. 2, Chile and Peru, 1865–1891, 30, 100; FO 16/181, Rumbold to Foreign Office, 12 April 1874; Farcau, *Ten Cents,* 17, 49–50; Rauch, *Southern Cone,* 144.

13. FO 16/201, Pakenham to Foreign Office, 27 May 1879; ibid., St. John to Foreign Office, 7 July 1879; FO 16/206, Pakenham to Layard, 12 March 1880; FO 78/3081, Layard to Salisbury, 19 Feb. 1880; FO 16/210, Bless Gana to Foreign Office, 3 June 1880.

14. FO 16/232, John Walsham to Granville, Berlin, 5 Aug. 1881; FO 16/232, George Annisley to Granville, Hamburg, 6 Aug. 1881.

15. FO 16/233, Henry James and Farrer Herschell to Granville, 1 Jan. 1883.

16. FO 16/234, Elswick to Granville, 15 Jan. 1881.

17. Ibid., G. Gara Almante, Peruvian Legation, to Julian Pauncefote, 8 Jan. 1881.

18. FO 16/234, G. Gara Almonte to Lord Tenterdon, 14 Jan. 1881.

19. Quoted in William F. Sater, *Chile and the United States: Empires in Conflict* (Athens: University of Georgia Press, 1990), 52.

20. Ibid.

21. Jürgen Schaefer, *Deutsche Militärhilfe an Südamerika* (Bertelsmann Universitätsverlag, 1974), 46–47.

22. ADM 231/14, South America. Coast Defences, etc., 1888; William F. Sater and Holger H. Herwig, *The Grand Illusion: The Prussianization of the Chilean Army* (Lincoln: University of Nebraska Press, 1999), 136; FO 16/259, Kennedy (Santiago) to Salisbury, 12 April 1890; Maurice Zeitlin, *The Civil War in Chile* (Princeton, NJ: Princeton University Press, 1984), 212.

23. William F. Sater, *Chile and the War of the Pacific* (Lincoln: University of Nebraska Press, 1986), 274.

24. ADM 231/17, French Fleet and Dockyards, 1890.

25. FO 16/260, Kennedy (Santiago) to Salisbury, 5 Sept. 1890.

26. *BDFA,* part I, Series D, Latin America, vol. 5, Peru 1871–1912, 227–228.

27. Ibid.

28. FO 16/260, Kennedy (Santiago) to Salisbury, 5 Sept. 1890.

29. Schaefer, *Deutsche,* 34; Sater and Herwig, *Grand Illusion,* 136–140.

30. FO 16/259, Kennedy (Santiago) to Salisbury, 9 May 1890; Sater and Herwig, *Grand Illusion,* 140; Schaefer, *Deutsche,* 35–36.

31. FO 16/260, Kennedy to Salisbury, 7 Sept. 1890.; ibid., Kennedy (Santiago) to Salisbury, 8 Oct. 1890; Zeitlin, *Civil War,* 85–86; Scheina, *Latin America's Wars,* 397.

32. *BDFA,* Chile and Peru, 245, 247.

33. Ibid., 253–254, 259–260, 280.

34. ADM 231/19, South America. Coast Defences, 1891.

35. Sater and Herwig, *Grand Illusion,* 133.

36. Nunn, "South American," 76–77.

37. Rauch, *Southern Cone,* 111, 114; Sater and Herwig, *Grand Illusion,* 144; ADM 231/27, Central and South American Republics. War Vessels and Torpedo Boats, 1897.

38. Schaefer, *Deutsche,* 40.

39. Sater and Herwig, *Grand Illusion,* 144–145; Schaefer, *Deutsche,* 39–44.

40. Sater and Herwig, *Grand Illusion,* 146.

41. ADM 231/27, Central and South American Republics. War Vessels and Torpedo Boats, 1897.

42. Quoted in Sater and Herwig, *Grand Illusion,* 140.

43. Sater and Herwig, *Grand Illusion,* 141–143; Schaefer, *Deutsche,* 72.

44. Zdenek Jindra, "Zur Entwicklung und Stellung der kanonenausfurh der Firma Friedrich Krupp/Essen, 1854–1912," in *Vierteljahrschrift für Sozial- und Wirtschafts-geschichte,* Beiheft 120 (Stuttgart: Franz Steiner Verlag, 1995) 971, 968; Rauch, *Southern Cone,* 114, 128n.79; ADM 231/27, Central and South American Republics. War Vessels and Torpedo Boats, 1897.

45. ADM 231/27, Central and South American Republics. War Vessels and Torpedo Boats, 1897.

46. Quoted in Rauch, *Southern Cone,* 184.

47. ADM 231/18, Argentine Republic. Naval Establishments, 1890; FO 6/490, Haggard (Buenos Aires) to Lansdowne, 13 May 1905, enclosure, Juan A. Martin, "Memoria del Departamento de Marina, 1904–1905"; ADM 231/27, Central and South American Republics. War Vessels and Torpedo Boats, 1897.

48. Victor Bulmer-Thomas, *The Economic History of Latin America since Independence* (Cambridge: Cambridge University Press, 1994), 103.

49. Schaefer, *Deutsche,* 47, 49–50, 65; Rauch, *Southern Cone,* xi–xii, 57, 67, 70; Sater and Herwig, *Grand Illusion,* 142.

50. Warren Schiff, "The Influence of the German Armed Forces and War Industry on Argentina, 1880–1914," *Hispanic American Historical Review* 52, no. 3 (1972): 438; Schaefer, *Deutsche,* 48; Rauch, *Southern Cone,* 184–185.

51. Lord to Hay, 10 May 1902, U.S. Department of State, *Despatches from US Ministers to Argentina,* Microcopy 69, Roll 35, vol. 40.

52. Ibid.
53. FO 466/667, Memorandum 28th May 1902, Argentina and Chile.
54. FO 46/667, Lowther (Santiago) to Lansdowne, 20 Aug. 1903.
55. Quoted in Marius B. Jansen, *The Making of Modern Japan* (Cambridge, MA: Belknap Press, 2000), 399–400.
56. Ibid., 427.
57. Richard J. Samuels, *"Rich Nation, Strong Army": National Security and the Technological Transformation of Japan* (Ithaca, NY: Cornell University Press, 1994), 88–89; Thomas L. Kennedy, *The Arms of Kiangnan: Modernization in the Chinese Ordnance Industry, 1860–1895* (Boulder, CO: Westview, 1978), 137.
58. ADM 231/2, no. 21, Capt. S. Long, "Japan. Report on Imperial Army and Navy," 17 March 1883; Kennedy, *Kiangnan,* 129; ADM 231/29, Japan. War Vessels and Torpedo Boats, 1898.
59. Jansen, *Making Japan,* 397; ADM 231/24, Japan. Coast Defences, Dockyards, 1894; ADM 231/2, no. 21, Capt. S. Long, "Japan. Report on Imperial Army and Navy," 17 March 1883; Jindra, "Entwicklung," 967, 971–972.
60. David C. Evans and Mark R. Peattie, *Kaigun: Strategy, Tactics and Technology in the Imperial Japanese Navy, 1887–1941* (Annapolis, MD: Naval Institute Press, 1997), 5–14.
61. ADM 231/10, Shipbuilding Policy of Foreign Nations, 1886, no. 119; ADM 231/7, French Fleet and Dockyards, 1885; ADM 231/11, French Fleet and Dockyards, 1887; ADM 231/12, French Fleet and Dockyards, 1888; Evans and Peattie, *Kaigun,* 15–16; ADM 231/29, Japan. War Vessels and Torpedo Boats, 1898; Roger Chesneau and Eugene Kolesnik, eds., *Conway's All the World's Fighting Ships, 1860–1905* (New York: Naval Institute Press, 1976), 216.
62. ADM 231/29, Japan. War Vessels and Torpedo Boats, 1898; Evans and Peattie, *Kaigun,* 17.
63. Jansen, *Making Japan,* 418, 421, 430, 439.
64. Quoted in Jansen, *Making Japan,* 432; John L. Rawlinson, *China's Struggle for Naval Development, 1839–1895* (Cambridge, MA: Harvard University Press, 1967), 144.
65. ADM 231/36, Naval Progress and Estimates, 1902.
66. WO 33/65, Reports on Changes in Various Foreign Armies during the Year 1896; WO 33/2818, Reports on Changes in Various Foreign Armies, 1897.
67. ADM 231/33, France. Guns, Torpedoes, 1900.
68. Udo Ratenhof, *Die Chinapolitek des Deutsches Reiches, 1871 bis 1945* (Boppard um Rhien: Harald Boldt Verlag, 1987), 218n.71; Jindra, "Entwickung," 969.

69. Evans and Peattie, *Kaigun,* 60; *Conway's, 1860–1905,* 217; ADM 231/29, Japan. War Vessels and Torpedo Boats, 1898.

70. ADM 231/32, Naval Estimates, 1900.

71. Evans and Peattie, *Kaigun,* 38, 62–64.

72. ADM 231/29, Japan. War Vessels and Torpedo Boats, 1898.

73. Evans and Peattie, *Kaigun,* 54.

74. L. G. Beskrovny, *The Russian Army and Fleet in the Nineteenth Century,* trans. Gordon E. Smith (Gulf Breeze, FL: Academic International Press, 1996), 314–315.

75. WO 106/6179, Reports on Changes in Various Foreign Armies during the Year 1902.

76. FO 46/667, Foreign Office to Macdonald (Tokyo), telegram 62, 27 Nov. 1903.

77. Ibid., Spring Rice (St. Petersburg), telegram 217 to Foreign Office, 28 Nov. 1903.

78. Ibid., Macdonald (Tokyo) telegram 12, 15 Oct. 1903.

79. Ibid., MacDonald (Tokyo) to Lansdowne, 5 Dec. 1903.

80. Ibid., Claude MacDonald (Tokyo) to Lansdowne, 6 Dec. 1903.

81. Ibid., MacDonald (Tokyo), decypher, 5 Dec. 1903.

82. Ibid., MacDonald to Lansdowne, 18 Dec. 1903.

83. Ibid., MacDonald to Lansdowne, 24 Dec. 1903.

84. Ibid., Consul General Keene (Genoa) to Lansdowne, 1 Jan. 1904.

6. The Dreadnought Races

1. Clive Trebilcock, *The Vickers Brothers: Armaments and Enterprise, 1854–1914* (London: Europa Publications, 1977), 120–121, 123.

2. FO 371/403, Haggard (Rio) to Grey, 14 April 1908, Brazil, Annual Report, 1907; FO 46/667, Kerr (Santiago) to Lansdowne, 26 Jan. 1905.

3. FO 371/201, W. Haggard, Brazil, General Report for the Year 1906.

4. FO 371/402, Cheetham (Rio) to Grey, 23 Aug. 1908; FO 371/12, Lowther (Rio) to Grey, 17 Jan. 1906.

5. Frederick M. Nunn, "The South American Military Tradition: Preprofessional Armies in Argentina, Chile, Peru, and Brazil," in Linda Alexander Rodriguez, ed., *Rank and Privilege: The Military and Society in Latin America* (Wilmington, DE: Scholarly Resources, 1994), 81.

6. FO 371/13, Buchanan to Tornquist, 5 Oct. 1906.

7. ADM 231/27, Central and South American Republics. War Vessels and Torpedo Boats, 1897.

8. FO 371/201, General Report on Brazil for the Year 1906, W. Haggard.

9. FO 371/13, Dering (Rio) to Grey, 25 July 1906; ibid., Barclay (Rio) to Grey, 9 Sept. 1906.

10. FO 371/5, Harford (Buenos Aires) to Grey, 30 July 1906; ibid., Haggard (Buenos Aires) to Grey, 9 Aug. 1906; ibid., Haggard to Grey, 3 Sept. 1906.

11. Ibid., Haggard (Buenos Aires) to Grey, 30 Sept. 1906.

12. FO 371/13, Buchanan (Rio) to Tornquist, 5 Oct. 1906.

13. Ibid., Windham Baring (Buenos Aires) to Alfred Mildmay, 12 Oct. 1906.

14. Ibid., Barclay (Rio) to Grey, 5 Oct. 1906.

15. FO 371/5, Haggard (Buenos Aires) to Grey, 30 Sept. 1906; FO 371/13 Foreign Office Minute, Brazil and Argentina: Naval Armaments, 28 Nov. 1906; ibid., Ernesto Tornquist to W. Buchanan, 28 Sept. 1906; FO 371/5 Haggard (Buenos Aires) to Grey, 10 Oct. 1906.

16. FO 371/13, O'Sullivan-Beare (Rio) to Grey, 10 Nov. 1906.

17. FO 371/12, Barclay (Rio) to Grey, 18 Nov. 1906.

18. FO 371/402, Cheetham (Rio) to Grey, 23 Aug. 1908; FO 371/13 Barclay (Rio) to Grey, 3 Dec. 1906; ibid., Colville Barclay to Grey, 4 Nov. 1906.

19. FO 371/194, Townley to Grey, 26 Dec. 1906.

20. FO 371/403, Haggard (Rio) to Grey, 17 May 1908.

21. Ibid., Cheetham (Rio) to Grey, 13 Oct. 1908; ibid., Capt. Hood to Grey, 3 Nov. 1908, Report on Navies of South America, 1908.

22. Ibid., Cheetham (Rio) to Grey, 27 Sept. 1908; ibid., Bertie (Paris) to Grey, 24 Sept. 1908.

23. Ibid., Milne Cheetham (Rio) to Grey, 21 July 1908.

24. Ibid., Townley (Rio) to Grey, 10 Aug. 1908.

25. Ibid., Henderson (Buenos Aires) to Grey, 12 Aug. 1908, enclosure, telegram from J. Percy Clark to Foreign Office, 10 Aug. 1908.

26. Ibid., J. Percy Clark (Buenos Aires) to Allen, telegram, 4 Aug. 1908.

27. Ibid., Allen to Clarke, 5 Aug. 1908.

28. Ibid., Henderson (Buenos Aires) to Grey, 14 Aug. 1908, enclosure, telegram, Clarke, Aug. 13, 1908.

29. Ibid., Cheetham (Rio) to Grey, 2 Aug. 1908; ibid., Cheetham to Grey, 18 Aug. 1908.

30. Ibid., Cheetham (Rio) to Grey, 23 Aug. 1908.

31. Ibid., Cheetham (Rio) to Grey, 9 Sept. 1908.

32. Ibid., Townley (Buenos Aires) to Grey, 28 Aug. 1908; ibid., Cheetham to Grey, 18 Aug. 1908; ibid., Cheetham (Rio) to Grey, 31 Aug. 1908; ibid., Russell (Buenos Aires) to Grey, 18 Dec. 1908.

33. FO 371/603, Haggard (Buenos Aires) to Grey, 22 Aug. 1909; FO 371/403, Capt. Hood to Grey, 3 Nov. 1908, Report on Navies of South America, 1908; ibid., Cheetham (Rio) to Grey, 12 Dec. 1908.

34. FO 371/403, Cheetham (Rio) to Grey, 27 Sept. 1908; ibid., Cheetham

(Rio) to Grey, 28 Sept. 1908; FO 371/603, Herbert Grant Watson to Grey, 16 Aug. 1909; FO 371/1518, Haggard (Rio) to Grey, 29 July 1913; FO 371/1580, Haggard (Rio) to Grey, 14 Sept. 1913.

35. FO 371/1051, Rear Admiral Farquhar to Admiralty, 12 Dec. 1910; ibid., Haggard (Rio) to Grey, 12 Dec. 1910; ibid., Haggard (Rio) to Grey, 13 Dec. 1910.
36. Ibid., Haggard (Rio) to Grey, 3 Feb. 1911.
37. Ibid., Haggard (Rio) to Grey, 22 Jan. 1911.
38. Ibid., Haggard (Rio) to Grey, 10 March 1911; ibid., Haggard to Grey, 19 March 1911; ibid., Haggard (Rio) to Grey, 1 April 1911; ibid., Haggard to Grey, 18 July 1911.
39. FO 371/1518, Haggard (Rio) to Grey, 19 June 1913, Brazil, Annual Report, 1912.
40. Ibid.
41. Ibid.
42. FO 371/1580, Haggard (Rio) to Grey, 19 Jan. 1913; ibid., Haggard to Grey, 4 Sept. 1913; ibid., Haggard to Grey, 6 Sept. 1913; ibid., Haggard to Grey, 9 Sept. 1913; ibid., Haggard (Rio) to Grey, 14 Sept. 1913.
43. FO 371/1518, Haggard (Rio) to Grey, 5 May, 1913; ibid., Grogan to Haggard, 7 June 1913; FO 371/1580, Robertson (Rio) to Foreign Office, 8 Dec. 1913; ibid., Robertson to Foreign Office, 12 Dec. 1913; ibid., Robertson to Foreign Office, 15 Dec. 1913.
44. FO 6/485, Haggard to Lansdowne, 13 Oct. 1904.
45. Ibid., Haggard (Buenos Aires) to Lansdowne, 24 Aug. 1904.
46. FO 6/490, Haggard (Buenos Aires) to Lansdowne, 13 May 1905.
47. FO 371/4, Harford (Buenos Aires) to Grey, 1 Jan. 1906, enclosed, Argentina, Annual Report, 1905; ibid., Harford to Grey, 21 Feb. 1906; ibid., Harford to Grey, 2 March 1906.; ibid., Harford to Grey, 1 April 1906.
48. GFM 21/215, Hacke to Bülow, 31 March 1906; FO 371/4, Harford to Grey, 29 May 1906; FO 371/194, Townley (Buenos Aires) to Grey, 11 April 1907.
49. FO 371/5, Haggard (Buenos Aires) to Grey, 5 Dec. 1906.
50. FO 371/200, Haggard (Buenos Aires) to Grey, 10 Feb. 1907; ibid., Haggard to Grey, 16 March 1907.
51. FO 371/194, Townley to Grey, 31 Jan. 1907.
52. Ibid., Townley (Buenos Aires) to Grey, 11 April 1907.
53. GFM 21/215, Waldthausen to Bülow, 28 Feb. 1908; ibid., Waldthausen to Bülow, 2 Feb. 1908; ibid., Waldthausen to Bülow, 26 March 1908.
54. FO 371/194, Townley (Buenos Aires) to Grey, 11 April 1907.
55. Ibid., Baird (Buenos Aires) to Grey, 4 Sept. 1907; FO 371/598, Claud Russell (Buenos Aires) to Grey, 31 Dec. 1908.
56. FO 371/397, Townley (Buenos Aires) to Grey, 29 Aug. 1908.

57. FO 371/824, Townley (Buenos Aires) to Grey, 19 Jan. 1910, Argentine Republic, Annual Report, 1909.
58. FO 371/397, Townley (Buenos Aires) to Grey, 29 Aug. 1908; GFM 21/215, Waldthausen to Bülow, 7 Sept. 1908.
59. FO 371/824, Townley (Buenos Aires) to Grey, 19 Jan. 1910, Argentine Republic, Annual Report, 1909.
60. Seward W. Livermore, "Battleship Diplomacy in South America: 1905–1925," *Journal of Modern History* 16, no. 1 (1944): 34–39.
61. FO 371/1573, British Consulate-General (Boston) to Foreign Office, 24 March 1913.
62. Ibid., Grogan to Tower, 5 April 1913; ibid., Tower to Grey, 17 Feb. 1913; ibid., Tower to Grey, 18 March 1913.
63. FO 371/1573, Tower (Buenos Aires) to Grey, 3 Jan. 1913, Argentine Republic, Annual Report, 1912.
64. Livermore, "Battleship Diplomacy," 45.
65. FO 371/17, Raikes (Santiago) to Grey, 2 Aug. 1906; FO 371/206, Ernest Rennie to Grey, 15 Feb. 1907, General Report on Chile for 1906.
66. FO 371/206, Rennie to Grey, 2 Aug. 1907; ibid., Rennie to Grey, 22 March 1907.
67. Ibid., Rennie to Grey, 23 Sept. 1907.
68. Livermore, "Battleship Diplomacy," 40; FO 371/611, Gosling (Santiago) to Grey, 2 May 1909; ibid., Gosling to Grey, 30 May 1909; ibid., Henry Lowther (Santiago) to Grey, 21 Oct. 1909, Chile, Annual Report, 1908; FO 371/841, Lowther (Santiago) to Grey, 12 July 1910; ibid., Lowther (Santiago) to Grey, 20 Nov. 1910; ibid., Lowther (Santiago) to Grey, 18 Jan. 1910, Chile. Annual Report, 1909.
69. FO 371/1309, Vaughan to Grey, 29 Feb. 1912, Chile, Annual Report, 1911.
70. FO 371/1588, Lowther (Santiago) to Grey, 28 Jan. 1913, Chile, Annual Report, 1912.
71. FO 371/1309, Grey to Vaughan, 29 April 1912.
72. ADM 231/43, Report 757, Reports on Foreign Naval Affairs, 1905 (vol. I), 139.
73. FO 32/751, Elliot to Lansdowne, 5 July 1904; ADM 231/43, Report 757, Reports on Foreign Naval Affairs, 1905 (vol. I), 139.
74. Ibid., Elliot to Lansdowne, 13 Aug. 1904; ibid., Elliot to Lansdowne, 17 Oct. 1904.
75. SA, FX 0050–02, Note Résumant la Question des Contre-Torpilleurs Grecs, 20 June 1905; ADM 231/43, Report 757, Reports on Foreign Naval Affairs, 1905 (vol. I), 140; RGVIA f. 2000, op. 1, d. 853, ll. 13–14.
76. FO 32/759, Elliot to Lansdowne, 6 Oct. 1905.
77. FO 371/81, Young to Grey, 7 Sept. 1906.

78. FO 371/464, Greece, Annual Report, 1907.

79. Ibid., Alban Young to Grey, 18 Aug. 1908.

80. Ibid., Elliot to Grey, 16 Dec. 1908.

81. FO 371/679, Elliot to Grey, 26 Nov. 1909; ibid., Wyndham to Grey, 17 Dec. 1909; ibid., Elliot to Grey, 17 Dec. 1909.

82. ADM 231/49, Report 841, Turkey, Greece and Roumania, War Vessels, 1908; Nejat Gülen, *Dünden Bugüne Bahriyemiz* (Istanbul: Kastas A. S. Yayinlari, 1988), 134–135.

83. KA, Generalstab. Militarattaches. Konstantinopel, kt. 49, Res. 313, 6 Dec. 1908.

84. SA, FX 0042–02, Lichtenberger to Schneider et Cie, 9 Oct. 1905; ibid., Contract, 22 Jan. 1906; KA, Generalstab. Militarattaches. Konstantinopel, kt. 46, Res. 23, 19 Feb. 1906.

85. FO 371/146, Col. Surtees to O'Conor, 26 Jan. 1906; KA, Generalstab. Militarattaches. Konstantinopel, kt. 48, Res. 30, 22 Feb., 1908; ibid., Res. 70, 20 April, 1908.

86. FO 371/146, O'Conor to Grey, 19 Feb. 1906.

87. Ibid.

88. FO 371/354, O'Conor to Grey, 12 Nov. 1907.

89. Andrea Filippo Saba, "L'attivatà dell' Ansaldo nell'Impero Ottomano," in Peter Hertner, ed., *Storia dell 'Ansaldo 3. Dai Bombrini ai Perrone, 1903–1914* (Bari: Guis. Laterza & Figli, 1996), 100.

90. FO 371/354, O'Conor to Grey, 30 July 1907; ibid., O'Conor to Grey, 7 Aug. 1907; ibid., Admiralty to Foreign Office, 10 Sept. 1907.

91. Ibid., Armstrong Co. to Sir Charles Hardinge, 25 Oct. 1907.

92. Ibid., Foreign Office to O'Conor, 3 Dec. 1907.

93. *Conway's All the World's Fighting Ships, 1906–1921* (London: Conway Maritime, 1985), 388.

94. FO 371/559, Djevad Bey to Grey, 14 Sept. 1908; ibid., Foreign Office to Admiralty, 17 Sept. 1908; ibid., Admiralty to Foreign Office, 7 Oct. 1908, enclosure no. 1, McKenna to Gamble.

95. Ibid., Lowther to Grey, 26 Sept. 1908, enclosure.

96. FO 371/560, Capt. Williamson to Lowther, 2 Nov. 1908.

97. FO 371/559, Lowther to Grey, 19 Nov. 1908.

98. Ibid., Lowther to Grey, 21 Nov. 1908.

99. Ibid., Foreign Office to Admiralty, 9 Dec. 1908.

100. KA, Generalstab, Militarattaches. Konstantinopel, kt. 51, Res. 231, 11 April 1909; ibid., Res. 339, 3 July 1909.

101. FO 371/679, Wyndham to Grey, 2 Dec. 1909.

102. *GP,* vol. 27 (Berlin: Deutsche Verlagsgesellschaft für Politik und Geschichte, 1925), 288.

103. FO 371/1000, Lowther to Grey, 15 Jan. 1910.
104. Ibid., Lowther to Grey, 16 Jan. 1910; ibid., Lowther to Grey, 31 May 1910; *GP,* 288–289, 293–294, 298–299.
105. FO 371/1000, Lowther to Foreign Office, 13 July 1910.
106. *GP,* 301–302.
107. *GP,* 303–305.
108. *GP,* 307.
109. FO 371/1245, Turkey, Annual Report, 1910.
110. FO 371/1013, Goschen (Berlin) to Grey, 16 Aug. 1910.
111. Ibid., Lowther to Grey, 13 Aug. 1910.
112. FO 368/231, Palmers Shipbuilding to Foreign Office, 12 Dec. 1908; ibid., Mallet to Lowther, 18 Dec. 1908.
113. AJP, Accession 1770, Hagley Museum and Library, Box 11, J. E. Mathews (Bethlehem Steel) in Constantinople to Leishman (Amercian Embassy, Rome), 26 June 1910.
114. Ibid., Leishman (Rome) to Schwab, 9 July 1910.
115. TWAS 130/1268, Armstrong Board Meeting Minutes, no. 3, Meeting of Directors, 16 Feb. 1911; ibid., Meeting of Managing Committee, 2 March 1911.
116. FO 371/1239, Lowther to Grey, 29 March 1911.
117. Ibid., Lord Furness to Edward Grey, 25 April 1911, enclosure 1, Talbot to Gowan, Constantinople, 10 April 1911.
118. Ibid., Admiralty to Foreign Office, 13 May 1911.
119. Ibid., Lowther to Grey, 14 April 1911.
120. FO 371/1240, Minutes, Turkish Naval Construction, 7 Sept. 1911, Admiralty Note.
121. AJP, Box 11, Leishman (Rome) to Schwab, 9 July 1910.
122. FO 371/1239, Lord Furness to Edward Grey, 25 April 1911, enclosure 1, Talbot to Gowan, Constantinople, 10 April 1911.
123. Ibid., Lord Furness to Edward Grey, 25 April 1911, enclosure 4, extract from *Lloyd Ottoman,* 9 April 1911.
124. Ibid., Lord Furness to Edward Grey, 25 April 1911, enclosure 3, Talbot to Gowan, Constantinople, 9 April 1911.
125. Ibid., Lord Furness to Edward Grey, 25 April 1911, enclosure 1, Talbot to Gowan, Constantinople, 10 April 1911.
126. TWAS 130/1268, Armstrong Board Meeting Minutes, no. 3, Meeting of Directors, 14 Sept. 1911.
127. Gülen, *Dünden,* 185–186; Trebilcock, *Vickers Brothers,* 121, 130.
128. A. S. Avetian, *Germanskii imperialism na blizhem vostoke* (Moscow: Izdatel'stvo Mezhdunarodye Otnosheniya, 1966), 116; Gülen, *Dünden,* 181, 186.

129. Djemal Pasha, *Memories of a Turkish Statesman—1913–1919* (New York: Arno Press, 1973), 83–84, 90–91.

130. Ibid., 116; British and Foreign State Papers, 1915, vol. 109, *Correspondence Respecting Events Leading to the Rupture of British Relations with Turkey*, ed. Edward C. Bleck (London, 1919), 671.

131. Avetian, *Germanskii,* 116; Djemal Pasha, *Memories,* 95, 102.

132. Djemal Pasha, *Memories,* 102.

133. FO 368/231, Commercial, Turkey, Lowther to Grey, 10 Nov. 1908.

134. Ibid., Commercial Dept. to Lowther, 24 Nov. 1908.

135. TWAS 130/1268, Armstrong Board Meeting Minutes, no. 3, Meeting of Directors, Westminster, 17 Feb. 1910.

136. Trebilcock, *Vickers Brothers,* 123–124; Zafer Toprak, *Türkiye'de "Milli Iktisat, 1908–1918* (Ankara: MAYA Matbaacilik-Yayincilik, 1982), 362; J. D. Scott, *Vickers: A History* (London: Weidenfeld and Nicolson, 1963), 85; Rafii-Sukru Suvla, "The Ottoman Debt, 1850–1939," in Charles Issawi, ed., *The Economic History of the Middle East, 1800–1914* (Chicago: University of Chicago Press, 1975), 106.

137. TWAS 130/1268, Board Meeting Minutes Armstrong, no. 3, Meeting of Directors, 16 Oct. 1913.

138. Ibid., Board Meeting Minutes Armstrong, no. 3, Meeting of Directors, 20 Nov. 1913; ibid., Meeting of Directors, 16 July 1914; Avetian, *Germanskii,* 110–111.

139. Djemal Pasha, *Memories,* 92.

140. Ibid., 94.

141. Trebilcock, *Vickers Brothers,* 125–127.

142. Djemal Pasha, *Memories,* 95.

143. Paul G. Halpern, *The Mediterranean Naval Situation, 1908–1914* (Cambridge, MA: Harvard University Press, 1971), 326.

144. Ibid., 343.

145. Henry Morgenthau, *Ambassador Morgenthau's Story* (New York: Doubleday, Page, 1918), 52–54; Halpern, *Naval Situation,* 351.

146. L. G. Beskrovnyi, *Russkaia armiia i flot v XIX veke* (Moskva: Nauka, 1973), 600; "Eksport Kazennyx Deneg za Granitsy," *Promyshlennost' i Torgovlia,* no. 24 (15 Dec. 1909): 641–644; *Promyshlennost' i Torgovlia* (1 Aug. 1908): 124–125; Peter Gatrell, *Government, Industry and Rearmament in Russia, 1900–1914* (Cambridge: Cambridge University Press, 1994), 276–77.

147. TWAS, Rendel Papers 31/7595–7618, Ottley to Falkner, 23 Sept. 1912.

148. Ibid., telegram from Ottley to London Office, 26 Sept. 1912.

149. Ibid., Ottley to Falkner, 23 Sept. 1912.

150. FO 371/1743, Buchanan to Grey, 2 Jan. 1913.

151. Ibid., Grenfell to Buchanan, 2 April 1913.

152. K. F. Shatsillo, "Inostrannyi Kapital i Voenno-morskie Programmy Rossii," *Istoricheskii Zapiski*, no. 69 (1961): 91–92.

153. J. N. Westwood, *Russian Naval Construction, 1905–45* (London: Macmillan, 1994), 92–94.

154. Scott, *Vickers*, 85, 134–135, 155; Trebilcock, *Vickers Brothers*, 135.

155. FO 371/1469, Grenfell to Buchanan, 2 Oct. 1912. See also Vladimir Vasil'evich Polikarpov, "Vikkers na Volge (1913–1917gg.)," *Voprosy Istorii*, no. 7 (1995): 121–132.

7. Gunning for Krupp

1. William F. Sater and Holger H. Herwig, *The Grand Illusion: The Prussianization of the Chilean Army* (Lincoln: University of Nebraska Press, 1999), 156; WO 106/6184, Report on Changes in Foreign Armies during 1907; ibid., Report on Changes in Foreign Armies during 1908; Robert W. Ball, *Mauser Military Rifles of the World* (Iola, WI: Krause Publications, 1996), 8; Wolfgang Seel, *Mause, von der Waffenschmiede zum Weltunternehmen* (Zürich: Verlag Stoker-Schmid AG, 1988), 64.

2. Claude Beaud, "Les Schneider marchands de Canons, 1870–1914," *Histoire, Economie et Societe* 14, no. 1 (1995): 127–129.

3. SAP, Protokoll uber die Verwaltungsrat-Sitzung, no. 14, 20 Aug. 1902; ibid., no. 17, 12 March 1903; ibid., no. 18, 18 July 1903; ibid., no. 21, 29 Jan. 1904; SAP, Geschäfts-Bericht, 1913, 2 Aug. 1913.

4. Rudolf Agstner, "Österreich-Ungarns Rüstungsexporte 1900–1914," *Österreichische Militärische Zeitschift* 35, no. 2 (1997): 169–170.

5. Udo Ratenhof, *Die Chinapolitik des Deutschen Reiches, 1871 bis 1945* (Boppard am Rhein: Harald Boldt Verlag, 1987), 221–222, 238.

6. Ibid., 218n71; Agstner, "Österreich-Ungarns," 170–171.

7. RGVIA f. 2000, op. 1, d. 772, l. 26; FO 78/5294, Elliot to Lansdowne, 10 June 1903; ibid., Elliot to Lansdowne, 14 July 1903; FO 78/5360, Buchanan to Lansdowne, 5 March 1904.

8. FO 78/5360, Marling to Lansdowne, 9 Jan. 1904.

9. Ibid., Massy to Consul General Sofia, 7 Jan. 1904.

10. SA, SS 0052–32, Revol to Geny, Sofia, 21 Feb. 1904; FO 78/5360, Massy to Marling, 4 March 1904.

11. FO 78/5360, Massy to Marling, 9 April 1904; ibid., Du Cane to Buchanan, 13 April 1904; ibid., Du Cane to Buchanan, 4 May 1904, enclosed in Buchanan to Lansdowne, 4 May 1904.

12. SA, SS 0424, Paris, 21 Sept. 1904.

13. SA 01G0560, folder 12, Report from Revol, Sofia, 30 Oct. 1904.

14. Ibid., folder 19, Report from Revol, Sofia, 8 Nov. 1904.

15. Ibid.; SA, FX 0037–18, Bulgarian contract, 13 Nov. 1904; FO 78/5361, Buchanan to Lansdowne, 14 Nov. 1904; ibid., Du Cane to Buchanan, 14 Nov. 1904.

16. FO 371/202, Akers-Douglas to Grey, 6 March 1907; ibid., Akers-Douglas to Grey, 27 Feb. 1907; ibid., Akers-Douglas to Grey, 22 March 1907.

17. RGVIA f. 2000, op. 1, d. 774, ll. 43–44; GFM 7/5, Romberg to Bülow, 31 March 1907.

18. GFM 7/5, Falkhausen to German Foreign Office, 11 March 1907.

19. FO 371/202, Akers-Douglas to Grey, 2 April 1907; ibid., Bulgaria, Annual Report, 1907.

20. Ibid., Du Cane to Buchanan, 18 Dec. 1907.

21. Ibid.

22. Ibid.

23. RGVIA f. 2000, op. 1, d. 774, l. 69.

24. TWAS, Rendel Papers 31/7616, Memorandum on Balkan States and Turkish Business (Lord Stuart Rendel), 31 Dec. 1912.

25. RGVIA f. 2000, op. 1, d. 774, l. 29.

26. GFM 7/5, General-Konsulat, Sofia, to Bülow, 21 Feb. 1906; ibid., Beckendorf to Bethmann-Hollweg, 14 Feb. 1910.

27. HHSA, PA, XIX Serbien, kt. 45, no. 1, telegram, 4 Jan. 1903; ibid., Gesandtschaftsarchiv. Belgrad, Telegraphen-Buch no. 9, telegram 1, 4 Jan. 1903; ibid., telegram 5, 9 Jan. 1903; ibid., telegram 32, 29 June 1903.

28. SAP, Protokoll über die Verwaltungsrat-Sitzung, Protokoll no. 18, 18 July 1903.

29. KA Generalstab. Militarattaches. Belgrad, kt. 6, Res. 89, Trappen to Pomiankowski, 13 May 1903.

30. KA Generalstab. Militarattaches. Belgrad, kt. 6, Res. 139, 30 July 1903; SAP, Protokoll no. 21, 29 Jan. 1904; Georg Günther, *Lebenserinnerungen* (Wien: Selbstverlage, 1936), 85; HHSA, PA XIX Serbien, kt. 45, no. 80 B, Dümba to Goluchkowski, 30 June 1903.

31. Wayne S. Vucinich, *Serbia between East and West: The Events of 1903–1908* (Stanford, CA: Stanford University Press, 1954), 189; SA 01G0560, folder 8, Business report, 27 Oct. 1904.

32. HHSA, Gesandtschaftsarchiv. Belgrad, Telegraphen-Buch no. 9, telegram 23, 3 Dec. 1904; ibid., telegram 18, 6 Dec. 1904; ibid., telegram 28, 7 Dec. 1904; ibid., telegram 19, 8 Dec. 1904; Vucinich, *Serbia*, 189.

33. SA 01G0560, folder 54, Business Report, 27 Dec. 1904.

34. RGVIA f. 2000, op. 1, d. 827, l. 11.

35. FO 371/130, Whitehead to Grey, 11 Dec. 1906, enclosure 1; Vucinich, *Serbia*, 190.

36. KA Generalstab. Militarattaches. Belgrad, kt. 7, Res. 62, 2 March, 1905; ibid., Res. 76, 17 March, 1905; ibid., Res. 89, 2 April, 1905.

37. HHSA, Gesandtschaftsarchiv. Belgrad, Telegraphen-Buch no. 9, telegram 18, 31 March 1905.

38. KA Militarattaches. Belgrad, kt. 7, Res. 108, 27 April 1905.

39. SA 01G0560, folder 86, Business Report, 20 Feb. 1905; ibid., folder 110, Business Report, 12 April 1905; ibid., folder 117, Business Report, May 1905; Vucinich, *Serbia,* 191.

40. FO 371/130, Whitehead to Grey, 11 Dec. 1906, enclosure 1; SAP, Protokoll no. 33, 16 Oct. 1905; HHSA, Gesandtschaftsarchiv. Belgrad, Telegraphen-Buch no. 9, telegram 55, 5 Nov. 1905; Vucinich, *Serbia,* 182–184, 192.

41. HHSA, Gesandtschaftsarchiv. Belgrad, Telegraphen-Buch no. 9, telegram 55, 22 March 1906; ibid., telegram 60, 4 April 1906.

42. Ibid., telegram 21, 6 April, 1906; ibid., telegram 62, 6 April 106; ibid., telegram 63, 6 April 1906; ibid., telegram 64, 7 April 1906.

43. Ibid., telegram 65, 8 April 1906.

44. Ibid., telegram 66, 9 April 1906.

45. Vucinich, *Serbia,* 192; FO 371/130, Thesiger to Grey, 19 April 1906.

46. HHSA, Gesandtschaftsarchiv. Belgrad, Telegraphen-Buch no. 9, telegram 73, 25 April 1906; ibid., telegram 81, 6 May 1906.

47. FO 371/130, Thesiger to Grey, 2 May 1906.

48. HHSA, Gesandtschaftsarchiv. Belgrad, Telegraphen-Buch no. 9, telegram 88, 22 May 1906; ibid., telegram 91, 31 May 1906; ibid., telegram 31, 11 June 1906; FO 371/130, Whitehead to Grey, 25 Aug. 1906; ibid., Whitehead to Grey, 11 Dec. 1906, enclosure 1.

49. HHSA, Gesandtschaftsarchiv. Belgrad, Telegraphen-Buch no. 9, telegram 32, 27 June 1906; HHSA, PA XIX, Serbien, kt. 53, No. 70l, 16 June 106; HHSA, Gesandtschaftsarchiv. Belgrad, Telegraphen-Buch no. 9, telegram 102, 7 July 1906; ibid., telegram 40, 26 July 1906; FO 371/130, Thesiger to Grey, 11 July 1906; ibid., Thesiger to Gey, 9 Aug. 1906.

50. FO 371/130, Whitehead to Grey, 25 Aug. 1906.

51. Ibid., Whitehead to Grey, 14 Nov. 1906; HHSA, Gesandtschaftsarchiv. Belgrad, Telegraphen-Buch no. 10, telegram 185, 13 Nov. 1906; ibid., telegram 186, 15 Nov. 1906.

52. FO 371/130, Whitehead to Grey, 11 Dec 1906, enclosure 1.

53. RGVIA f. 2000, op. 1, d. 823, chast' 2, l. 9; FO 371/130, Whitehead to Grey, 20 Dec. 1906; HHSA, Gesandtschaftsarchiv. Belgrad, Telegraphen-Buch no. 10, telegram 192, 10 Dec. 1906; ibid., telegram 194, 10 Dec. 1906.

54. HHSA, Gesandtschaftsarchiv. Belgrad, Telegraphen-Buch no. 10, telegram 9, 22 April, 1907; ibid., telegram 12, 29 April 1907; ibid., Telegram 13, 2

May 1907; ibid., telegram 21, 25 July 1905; FO 371/328, George Young to Grey, 4 May 1907.

55. Constantin Dumba, *Dreibund- und Entente-Politik in der Alten und Neuen Welt* (Zurich: Amalthea-Verlag, 1931), 219; Vucinich, *Serbia*, 195.

56. SAP, Protokoll no. 19, 22 Oct. 1903; ibid., Protokoll no. 21, 29 Jan. 1904; ibid., Protokoll no. 29, 26 Nov. 1904; SAP, Geschäfts-Bericht des Verwaltungsrathes der Skodawerke 1903/04; SAP, Protokoll no. 28, 3 Nov. 1904; ibid., Protokoll no. 33, 16 Oct. 1905.

57. HHSA, PA XIX, Serbien, kt. 48, no. 2, Baron Flotow to Goluchowski, 6 Jan. 1904; Dumba, *Dreibund*, 233.

58. FO 371/130, Whitehead to Grey, 14 Nov. 1906; Vucinich, *Serbia*, 190.

59. FO 371/130, Whitehead to Grey, 11 Dec. 1906, enclosure 1.

60. FO 371/328, General Report on the Kingdom of Serbia for the Year 1906.

61. WO 106/6185, Report on Changes in Foreign Armies during 1908; HHSA PA XII, Turkei, kt. 362 Liasses XXXIX/7, no. 1153, Report to Reichskriegsministerium, 25 March 1909; HHSA, PA XII, Turkei, kt. 362 Liasse XXXIX/7, no. 67B, London to Aehrenthal, 15 Dec. 1908; HHSA, PA XII, Turkei, kt. 362 Liasse XXXIX/7, Report no. 2837, 19 Dec. 1908; ibid., no. 69D, Bukarest, Schönburg to Aehrenthal, 19 Dec. 1908; ibid., no. 62B, Von Szogyeny to Aehrenthal, 22 Dec. 1908; RGVIA f. 2000, op. 1, d. 823, chast' 2, l. 81; HHSA PA XII, Turkei, kt. 362 Liasses XXXIX/7, no. 13, Belgrad, 6 Feb. 1909.

62. HHSA, PA XII, Turkei, kt. 362 Liasse XXXIX/7, Pras. no. 9399, 20 Oct. 1908; ibid., no. 1, Herr Hickel, Marseilles, 24 Oct. 1908; ibid., no. 2, Herr Hickel, Marseilles, 27 Oct. 1908.

63. Ibid., no. 457, Pallavacini, 3 Nov. 1908.

64. Ibid., Abschrift eines Privatbriefes aus Konstantinopel, 10 March 1909.

65. Ibid., no. 13D, Belgrad, 6 Feb. 1909; no. 60, Graf D. Thurn, 17 March 1909; ibid., no. 63, Graf Thurn, 20 March 1909; ibid., no. 119, Pallavicini to War Ministry, 9 March 1909; ibid., no. 1301, Pallavicini to War Ministry, 31 March, 1909.

66. FO 371/1472, Lt. Col. Lyon to Paget, 20 Sept 1912; ibid., Paget to Grey, 17 Sept. 1912; ibid., Grey to Paget, 16 Sept. 1912.

67. RGVIA f. 2000, op. 1, d. 827, l. 25; Raymond Poidevin, *Les Relations économiques et financiers entre la France et L'Allemagne de 1898 a 1914* (Paris: Armand Colin, 1969), 573–574; FO 371/1472, Ralph Paget (Belgrade) to Grey, 16 Sept. 1912.

68. SA, 01G0560, folder 11, note, 31 Oct. 1904; FO 32/759, Elliot to Lansdowne, 19 May 1905; FO 371/81, Alban Young to Grey, 24 July 1906.

69. RGVIA f. 2000, op. 1, d. 853, l. 106.

70. FO 371/464, Greece, Annual Report, 1907.

71. Ibid., Elliot to Grey, 7 Jan. 1908.
72. SA, SS 0426, P. Laurens to Schneider Co., 13 Aug. 1907.
73. Grèce, Artillerie de Campagne et de Montagne, *Rapport de la Majorité de la Commission,* 1907; RGVIA f. 2000, op. 1, d. 853, l. 106; KAE, WA 4/1281, telegram, Krupp to Greek War Ministry, 15 July 1907; ibid., Krupp to Greek War Ministry, 27 July 1907.
74. SA, SS 0426, Maurice Devies to Schneider Co., 16 July 1907, enclosure.
75. FO 371/679, Elliot to Grey, 25 Sept. 1909; ibid., Armstrong chairman to Secretary of State, Foreign Office, 14 Oct. 1909.
76. Ibid., Armstrong chairman to Secretary of State, Foreign Office, 14 Oct. 1909.
77. Ibid., Foreign Office to Armstrong, 21 Oct. 1909.
78. FO 371/679, Elliot to Grey, 8 Oct. 1909.
79. Ibid., Surtees to Minister Plenipotentiary, Athens (Sir Francis Smith), 8 Oct. 1909.
80. Baron Wladimir Giesl von Gieslingen, *Zwei Jahrzehnte im nahen Orient* (Berlin: Mittler & Sohn, 1929), 170.
81. RGVIA f. 2000, op. 1, d. 794, l. 17.
82. FO 104/155, Kennedy to Lansdowne, 30 June 1903.
83. Ibid., Kennedy to Lansdowne, 4 Feb. 1903.
84. Ibid., Kennedy to Lansdowne, 30 June 1903; RGVIA f. 2000, op. 1, d. 794, ll. 43–44, 47–48, 50, 59.
85. RGVIA f. 2000, op. 1, d. 794, ll. 44–45.
86. GFM 7/22, Rheinische Metallwaren to Aus. amt, 6 May 1905; ibid., Rheinische Metall to Aus. amt., 18 Jan. 1906; ibid., Von Hammerstein to war minister, 17 Dec. 1905.
87. Ibid., Von Hammerstein to war minister, 17 Dec. 1905.
88. Ibid., Von Hammerstein to War Ministry, 6 Jan. 1906.
89. Ibid., Von Hammerstein to embassy in Romania, 7 July 1906.
90. RGVIA f. 2000, op. 1, d. 794, l. 114.
91. FO 371/510, Greene to Grey, 22 Jan. 1908.
92. FO 371/511, C. Greene to Grey, 10 Feb. 1908.
93. GFM 7/22, Military attaché report, Romania no. 23, 24 June 1910.
94. RGVIA f. 2000, op. 1, d. 794, l. 116.
95. FO 371/1212, Arnold Robertson to Grey, 24 Feb. 1911.
96. Ibid.
97. GFM 7/22, Von Bronsart to Aus. amt., 26 Feb. 1912.
98. Ibid., Von Bronsart to Aus. amt, 4 March 1912; ibid., Rosen to Bethman-Hollweg, 10 March 1912.
99. Poidevin, *Les Relations,* 269–272; SA, 01G0560, folder 54, Turquie, 27 Dec. 1904.

100. Poidevin, *Les Relations,* 272; SA, 01G0560, folder 99, Turquie, 22 March 1905.
101. Poidevin, *Les Relations,* 273.
102. SA, 01G0560, folder 106, Turquie, 1 April 1905; KA, Generalstab. Militarattaches. Konstantinopel, kt. 45, Res. 28, 7 Feb. 1905.
103. Poidevin, *Les Relations,* 275.
104. WO 106/6182, Reports on Changes in Various Foreign Armies during the Year 1905; A. S. Avetian, *Germanskii imperializm na blizhnem vostoke* (Moskva: Izdatel'stvo Mezhdunarodye Otnosheniya, 1966), 109–110; Rafii-Sukru Suvla, "The Ottoman Debt, 1850–1939," in Charles Issawi, ed., *The Economic History of the Middle East, 1800–1914* (Chicago: University of Chicago Press, 1975), 105.
105. FO 371/348, Letter, Sir. W. G. Armstrong, Chairman of Armstrong, Whitworth to Sir Charles Hardinge, Under Secretary of State for Foreign Affairs, 25 March 1907.
106. Bernd Langensiepen and Ahmet Güleryüz, *The Ottoman Steam Navy, 1828–1923,* ed. and trans. James Cooper (Annapolis, MD: Naval Institute Press, 1995), 14.
107. KA, Generalstab. Militarattaches. Konstantinopel, kt. 49, Res. 266, Giesl to Conrad, 5 Nov. 1908; ibid., Res. 338, Giesl to Conrad, 19 Dec. 1908.
108. FO 371/560, Col. Surtees to Lowther, 23 Nov. 1908.
109. Ibid., Lowther to Grey, 8 Dec. 1908.
110. Ibid., Col. Surtees to Lowther, 23 Nov. 1908.
111. KA, Generalstab. Militarattaches. Konstantinopel, kt. 44, Res. 164, 9 Aug. 1904; ibid., Res. 182, 30 Aug. 1904.
112. FO 371/560, Col. Surtees to Lowther, 23 Nov. 1908.
113. Ibid.
114. FO 371/560, Col. Surtees to Lowther, 3 Dec. 1908.
115. FO 371/561, Lowther to Grey, 20 Dec. 1908, enclosure 1.
116. KA, Generalstab. Militarattaches. Konstantinopel, kt. 48, Res. 135, 16 June 1908.
117. FO 371/560, Falkner to Grey, 21 Dec. 1908.
118. Ibid., Lowther to Grey, 8 Dec. 1908.
119. KA, Generalstab. Militarattaches. Konstantinopel, kt. 51, Res. 344, 8 July 1909.
120. KA, Generalstab. Militarattaches. Konstantinopel, kt. 52, Res. 447, 26 Oct. 1909.
121. SA, 01G0906–01; SAP, Protokoll, 2 Aug. 1913; FO 371/1811, Lt. Col. Tyrrell to Lowther, 2 June 1913.
122. Djemal Pasha, *Memories of a Turkish Statesman—1913–1919* (New York: Arno, 1973), 102.

123. FO 371/194, Townley (Buenos Aires) to Grey, 2 May 1907; GFM 21/215, Krupp to Richtofen, 4 July 1903; FO 371/4, Harford (Buenos Aires) to Grey, 17 Jan. 1906; ibid., Harford to Grey, 29 May 1906.

124. GFM 21/215, RheinMetall to Aus. amt. 2 March 1907; ibid., Krupp to Bülow, 6 Nov. 1907.

125. FO 371/397, Townley to Grey, Report on Argentina for the Year 1907, 6 Feb. 1908.

126. Ibid., Townley (Buenos Aires) to Grey, 3 Aug. 1908.

127. FO 371/598, Russell (Buenos Aires) to Grey, 31 Dec. 1908.

128. FO 371/1295, Tower (Buenos Aires) to Grey, Argentine Republic, Annual Report, 1911, 15 Jan. 1912.

129. FO 371/598, Russell (Buenos Aires) to Grey, 31 Dec. 1908.

130. FO 371/206, Ernest Rennie to Grey, General Report on Chile for 1906, 15 Feb. 1907; Sater and Herwig, *Grand Illusion*, 147–148.

131. Sater and Herwig, *Grand Illusion*, 151, 155–156; FO 371/1588, Lowther (Santiago) to Grey, 12 Jan. 1913, enclosed, military report.

132. Sater and Herwig, *Grand Illusion*, 150, 162–163; FO 371/206, Rennie to Grey, 1 July 1907; FO 371/611, Gosling (Santiago) to Grey, 3 April 1909; ibid., Gosling to Grey, 2 May 1909; ibid., Gosling to Grey, 30 May 1909; FO 371/841, Lowther (Santiago) to Grey, Chile, Annual Report, 1909, 18 Jan. 1910.

133. FO 371/1588, Lowther to Grey, Chile, Annual Report, 1912, 28 Jan. 1913; FO 371/1309, Lowther (Santiago) to Grey, 11 Oct. 1912; ibid., Lowther (Santiago) to Grey, 13 Nov. 1912.

134. FO 371/1309, Lowther (Santiago) to Grey, 13 Nov. 1912.

135. Charles A. E. X. Maitrot, *La France et les republiques sud-americaines* (Nancy: Berger-Levrault, 1920), 20–23, 252; William Manchester, *The Arms of Krupp, 1587–1968* (Boston: Bantam Books, 1968), 305.

136. David Stevenson, *Armaments and the Coming of War: Europe, 1904–1914* (Oxford: Clarendon Press, 1996), 355.

Conclusion

1. *100 Jahre Schichau, 1837–1937* (Berlin: VDI-Verlag, 1937), 32–33, 38, 71–72; Eberhard Westpahl, *Ein Ostdeutscher Industriepionier: Ferdinand Schichau in sienem Leben und Schaffen* (Essen: West Verlag, 1957), 74–79.

2. James L. Payne, *Why Nations Arm* (Oxford: Blackwell, 1989), 2–3, 42–45.

3. Richard J. Samuels, *"Rich Nation, Strong Army": National Security and the Technological Transformation of Japan* (Ithaca, NY: Cornell University Press, 1994), 93; ADM 231/48, Foreign Naval Progress and Estimates, 1906–1907, 148; David C. Evans and Mark R. Peattie, *Kaigun: Strategy,*

Tactics, and Technology in the Imperial Japanese Navy, 1887–1941 (Annapolis, MD: Naval Institute Press, 1997), 63; ADM 231/48, Foreign Naval Progress and Estimates, 1906–1907, 175.

4. David Stevenson, *Armaments and the Coming of War: Europe, 1904–1914* (Oxford: Clarendon Press, 1996), 413; David G. Herrmann, *The Arming of Europe and the Making of the First World War* (Princeton, NJ: Princeton University Press, 1996), 130–136.

Index